Ted is my friend

A mathematics activity book

Betty Coombs Lalie Harcourt

Illustrated by
Steve Pilcher

Addison-Wesley Publishers

Don Mills, Ontario • Reading, Massachusetts
Menlo Park, California • Wokingham, Berkshire
Amsterdam • Sydney • Singapore • Tokyo
Madrid • Bogota • Santiago • San Juan

Dear friend

Hello! How are you? I am fine. My name is Ted. I am a Troll. Troll – Teddy is my pet. We live in a little white cabin in the forest. We are very happy here, but sometimes funny things happen. Yesterday, my cookie jar disappeared. Today I found it under my pillow and there were only two cookies left. Who do you think could have taken the others?

LOVE

Ted and troll-teddy.

P.S. SSshh! Ted just fell asleep. Now it's our turn to say hello. We are imps. There are lots and lots of us. We live with Ted but he doesn't know it. We play funny tricks on Ted. Ssh! Don't tell Ted, but we ate the cookies in his cookie jar!

the imps.

Ted is my Friend

Have the child draw a picture of herself or himself with Ted. Discuss how Ted and the child are alike; how they are different.

Ted and I

Ted has blue eyes.

I have __________ eyes.

Ted has pink hair.

I have __________ hair.

Ted has Troll Teddy.

I have __________.

Read the information with the child. Have the child complete the sentence by drawing.

My Ted Puppet

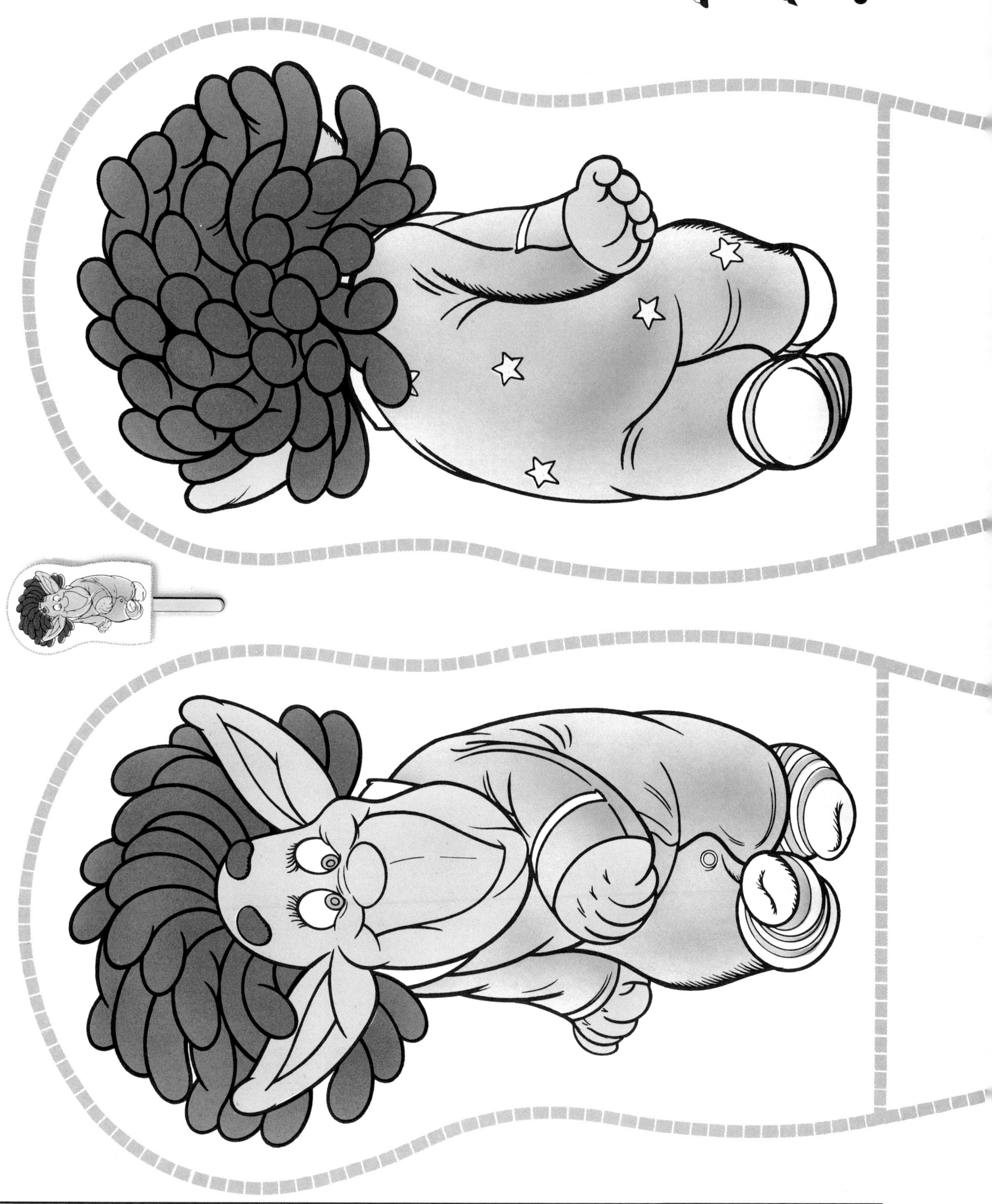

Have the child cut out the puppet on the dotted lines and glue back and front together on a popsicle stick. See page 93 for further direction.

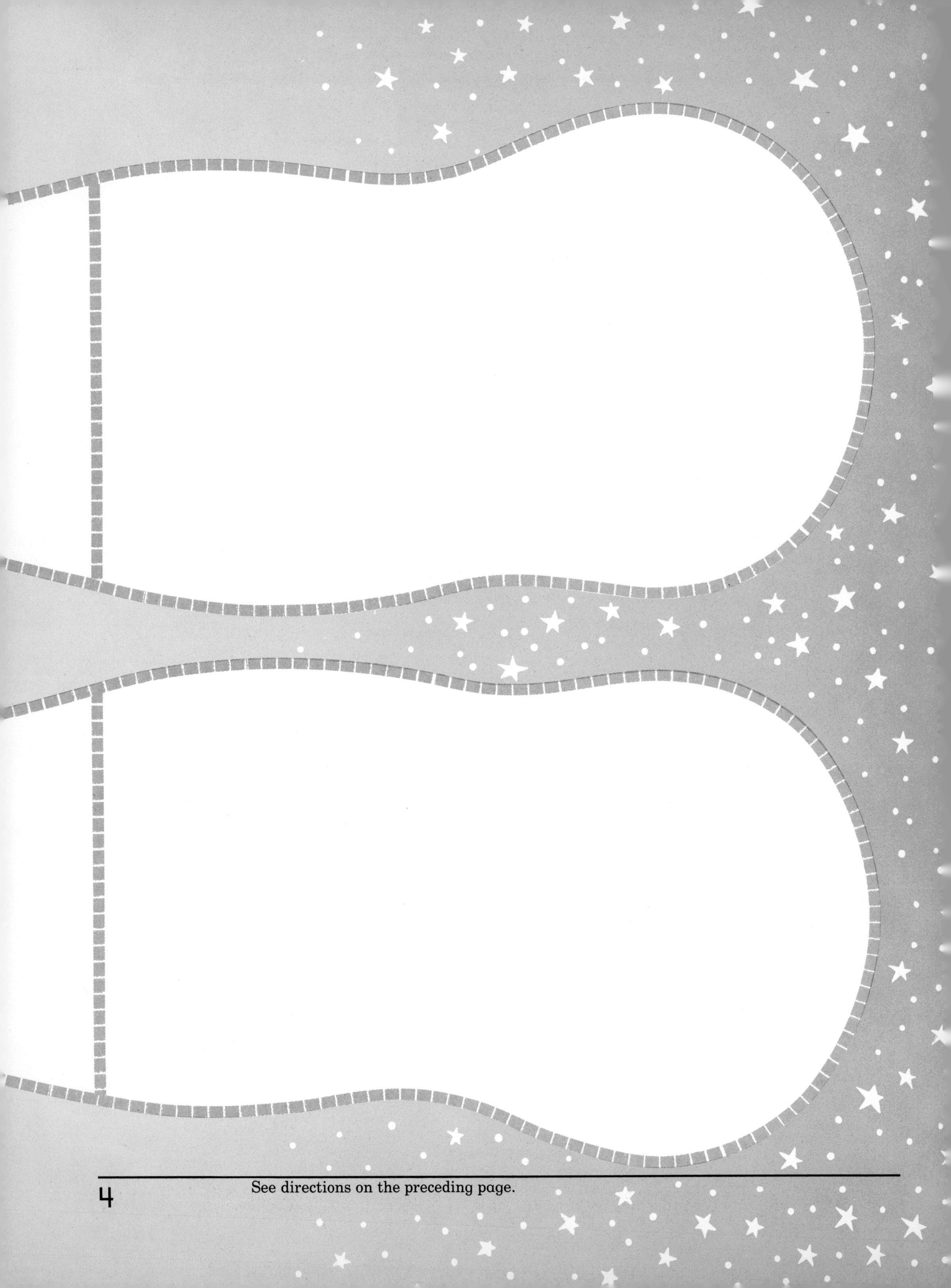

See directions on the preceding page.

My Troll-teddy Puppet

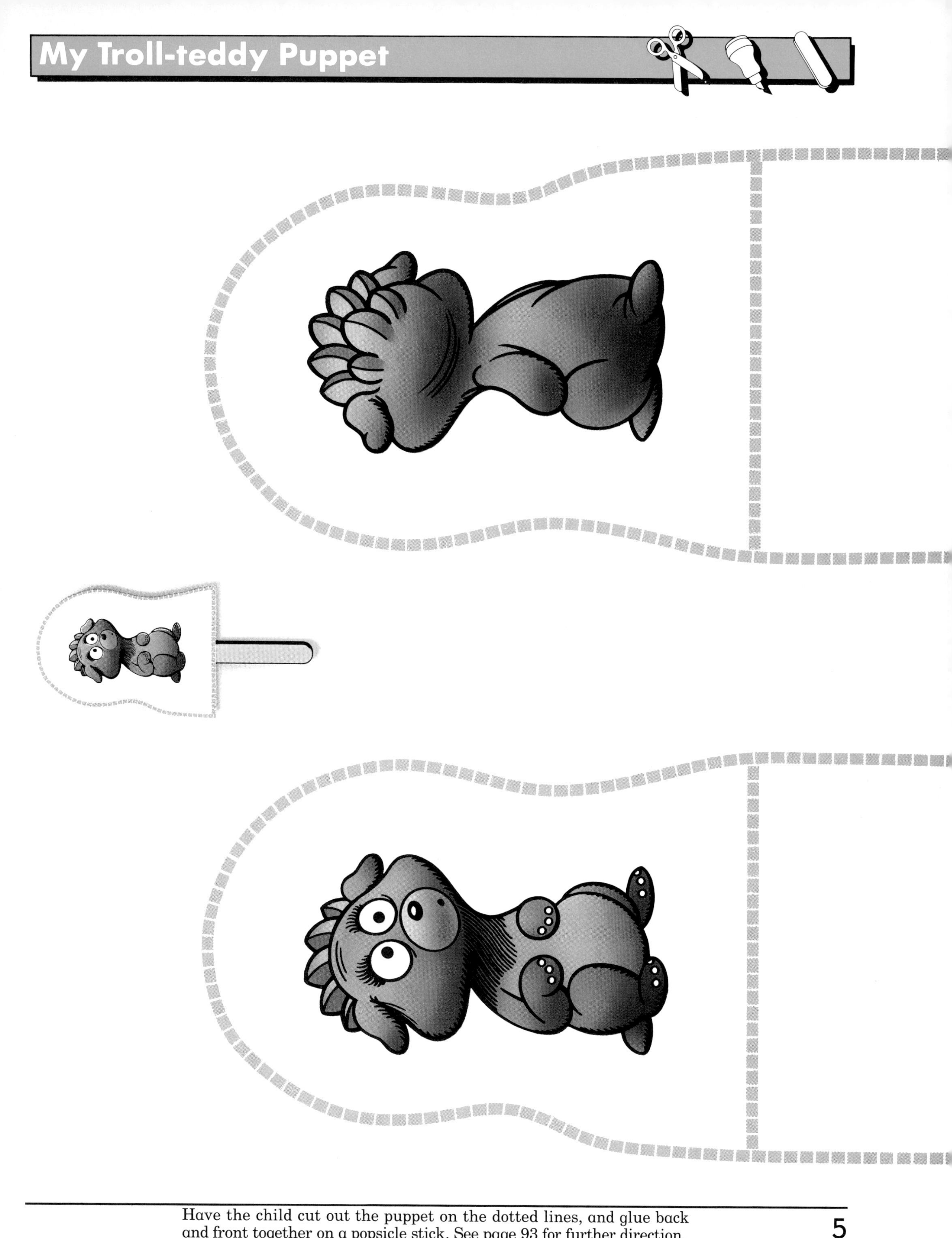

Have the child cut out the puppet on the dotted lines, and glue back and front together on a popsicle stick. See page 93 for further direction.

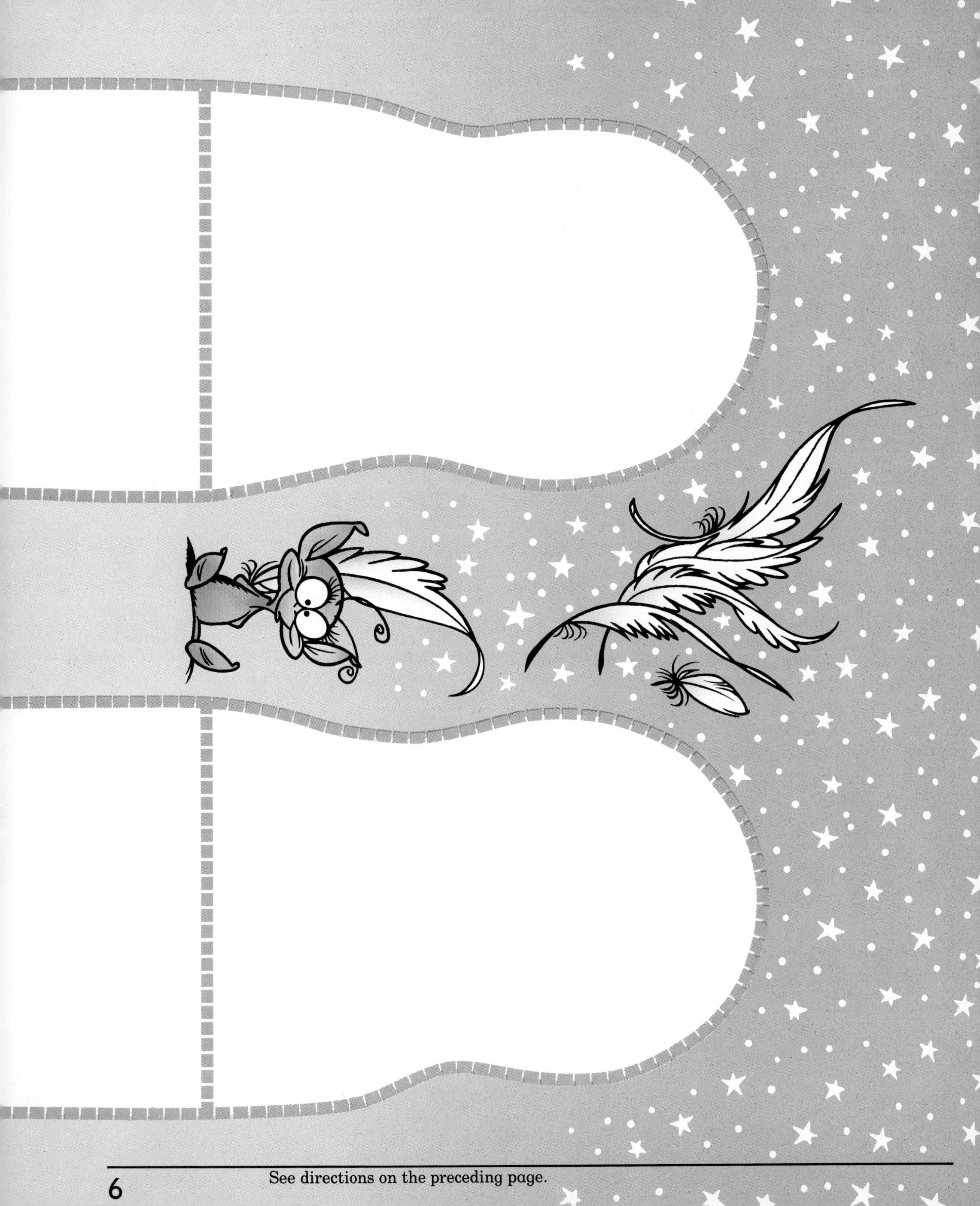

See directions on the preceding page.

My Imp Puppets

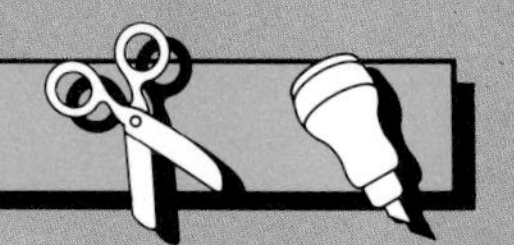

Kalaloo

Ari

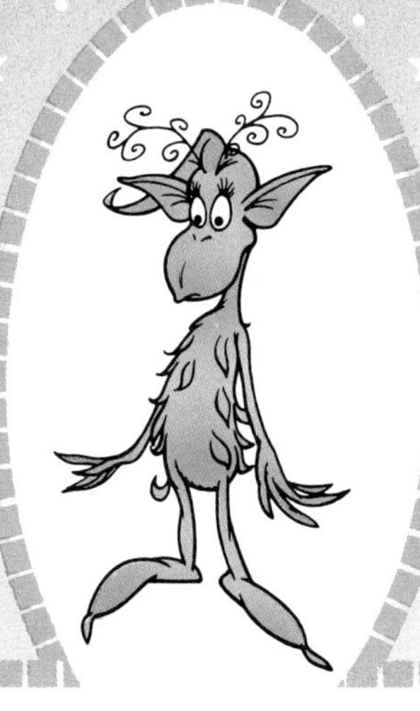

Rumple

Slider

Have the child cut out the imps. Roll and glue or tape the tabs to make finger puppets. See page 93 for further direction.

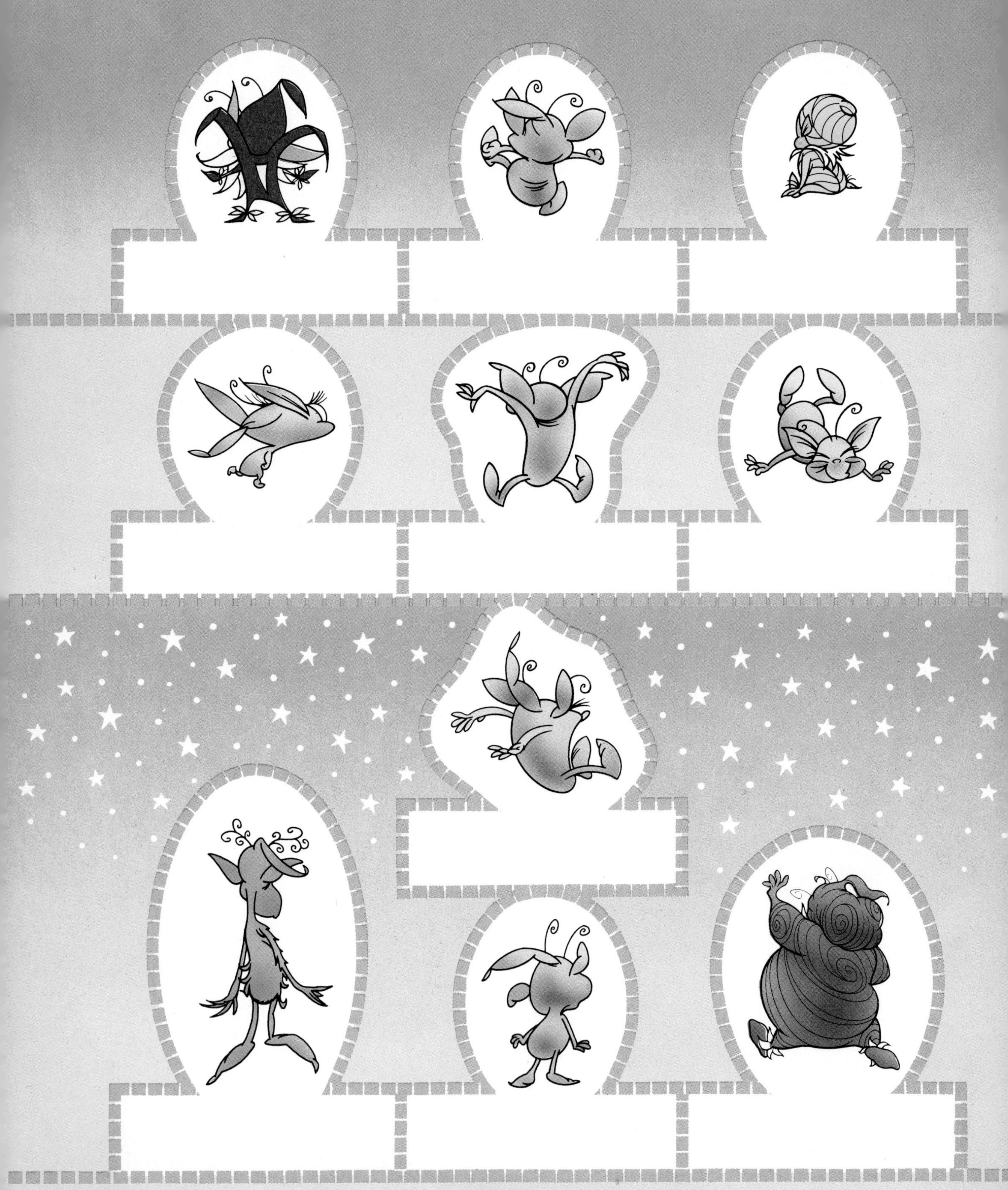

See directions on the preceding page.

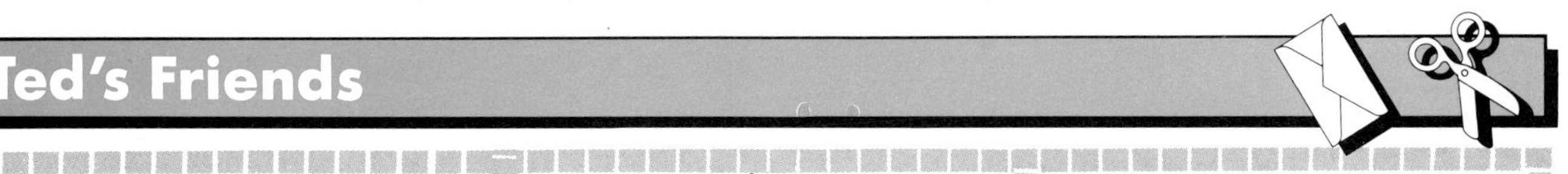

Ted's Friends

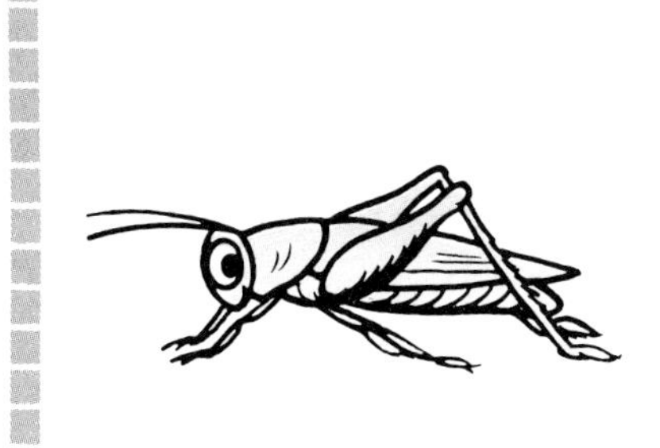

Have the child cut out the pictures. You may wish to store these in an envelope. See page 93 for further directions.

Ted's Friends

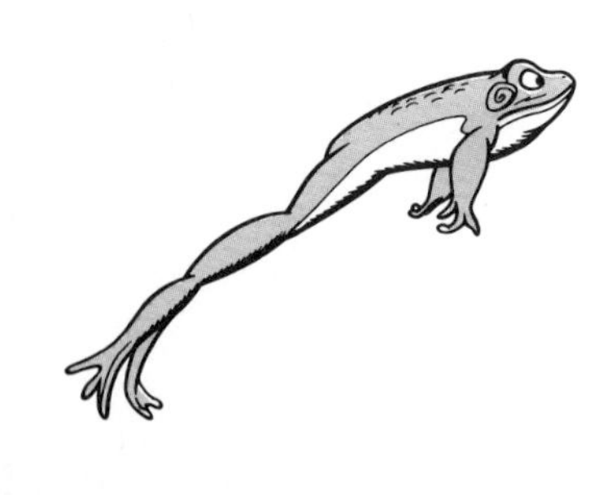

See directions on the preceding page.

Things from Ted's Cupboard

Have the child cut on the lines and sort the things onto the shelves on page 13. You may wish to store these in an envelope.

More Things from Ted's Cupboard

See directions on the preceding page.

Ted's Cupboard

Have the child sort Ted's things (see pages 11 and 12) onto his cupboard shelves in many different ways. The child may wish to glue them in the way he or she likes best.

Straw Patterns

Have the child use cut up straws to copy and then extend the patterns.
Encourage the child to describe each pattern.

A Necklace for my Friend

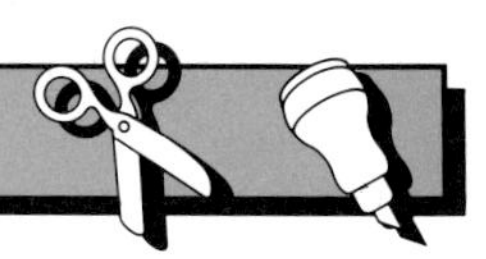

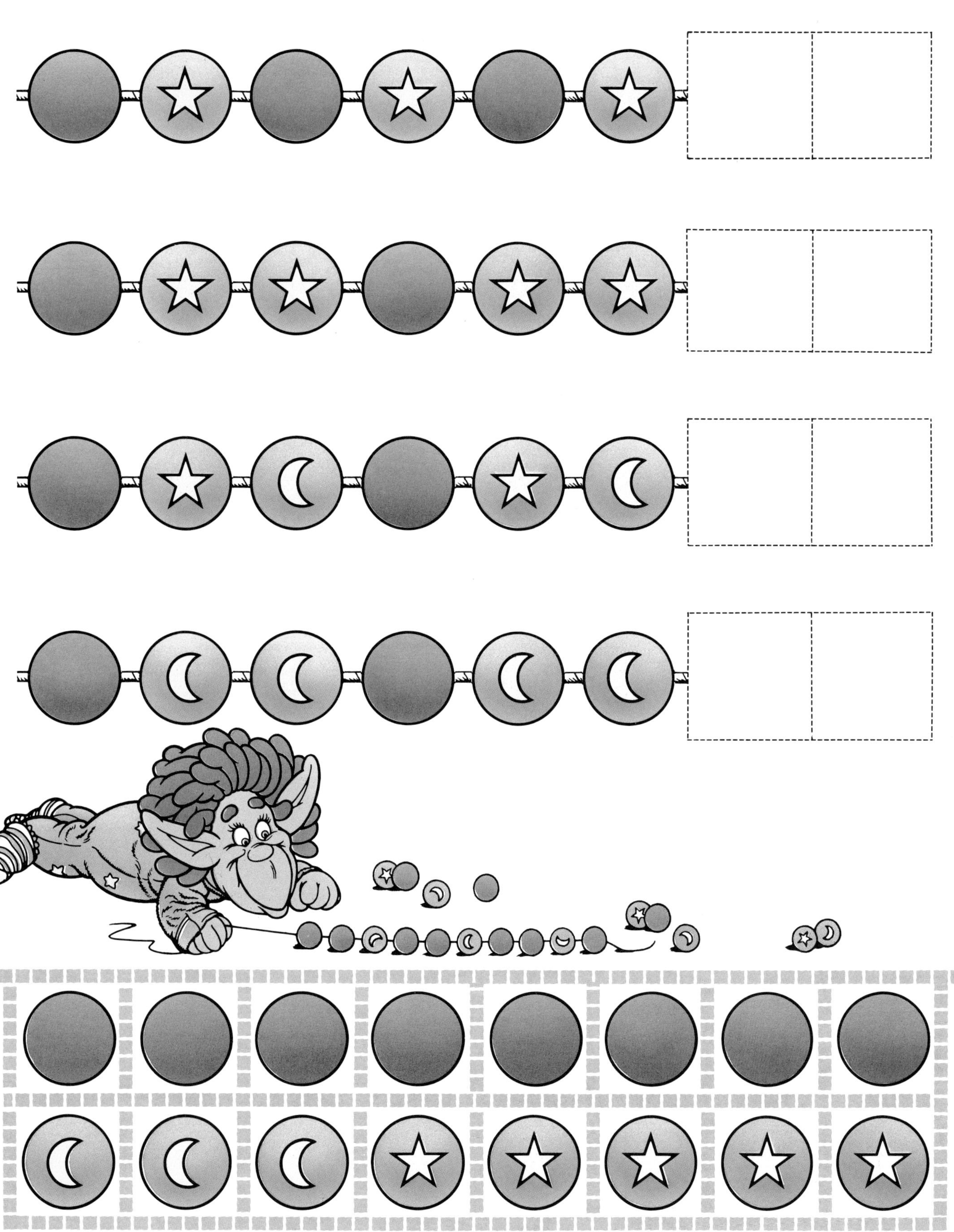

Have the child cut out the beads and finish the patterns. The child may wish to extend a pattern off the page. Encourage the child to describe the patterns.

Ted's Scarf

Have the child color Ted's scarf in a pattern. Encourage the child to describe the pattern.

Have the child color Ted's scarf in a pattern. Encourage the child to describe the pattern.

Ted's Quilt

Have the child cut out the squares to make a patterned quilt for Ted on page 19.

Ted's Quilt

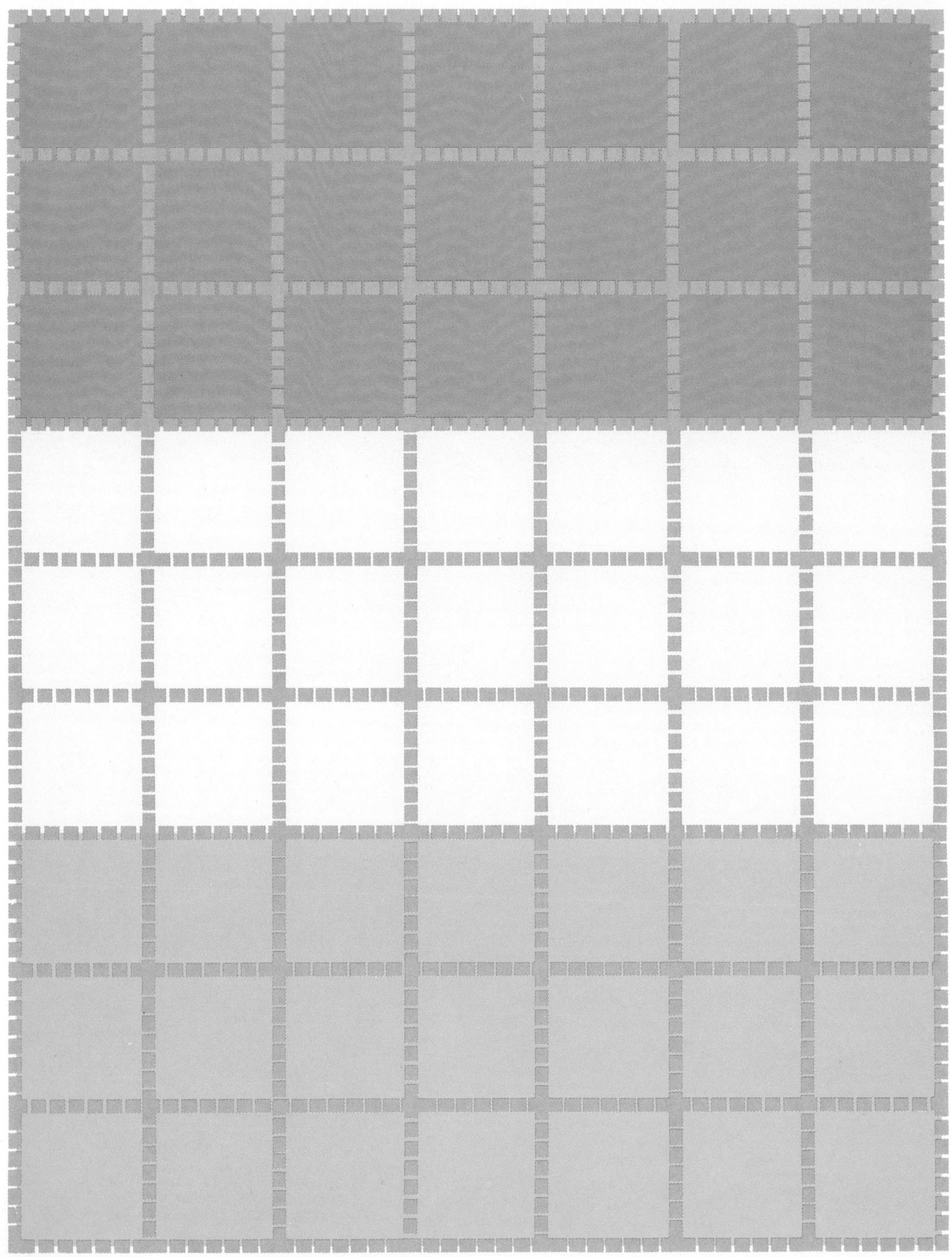

See directions on the preceding page.

Ted's Quilt

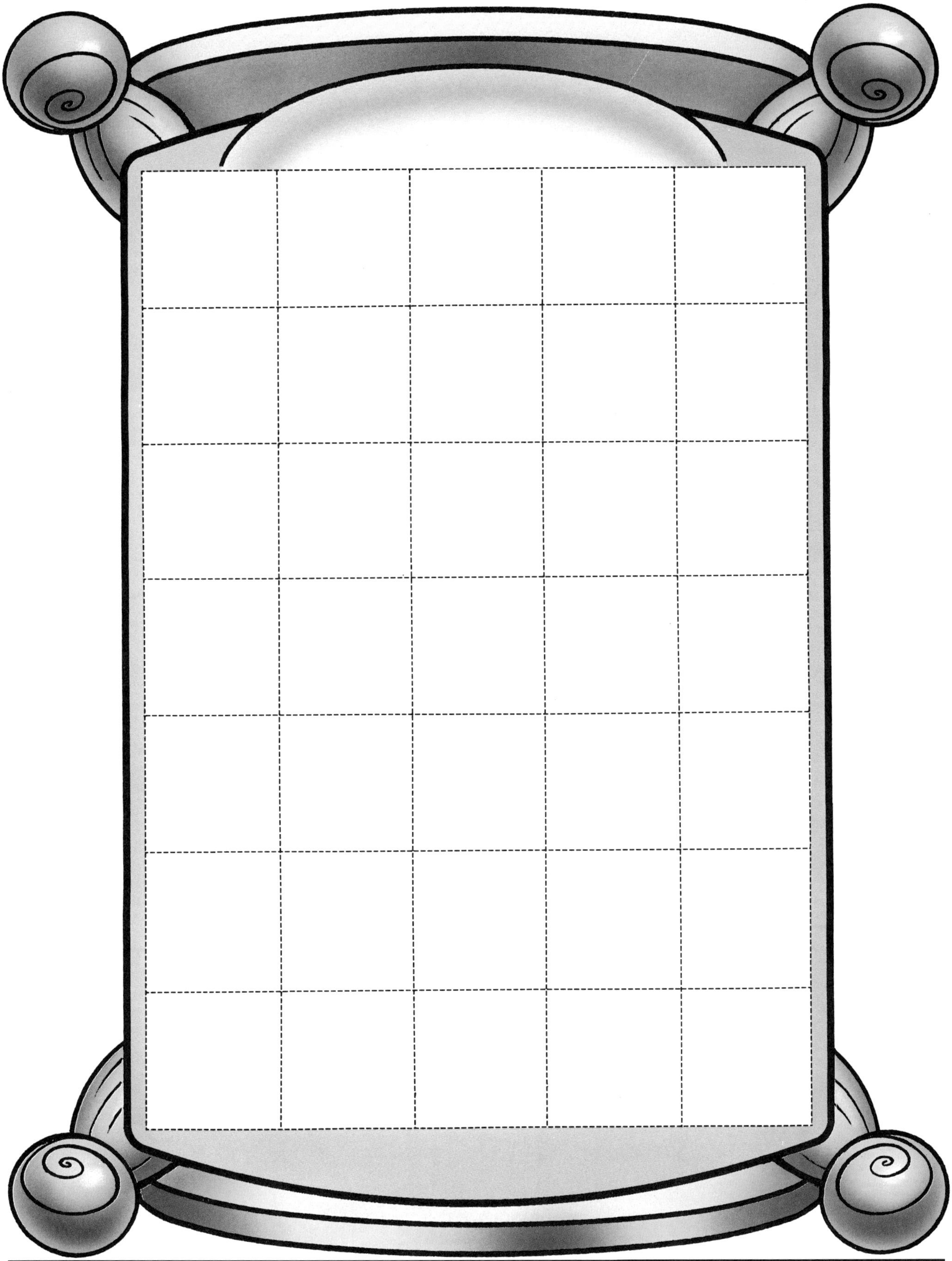

Have the child use the squares from pages 17 and 18 to make a patterned quilt. Encourage the child to describe her or his pattern.

Ted's Bug Collection

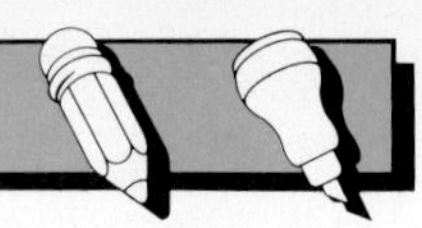

Have the child arrange sets of bugs from page 21 in Ted's jars and record how many are in each jar.

Ted's Bug Collection

Have the child cut and paste the bugs into Ted's jars, then record the number in each.

My Collection

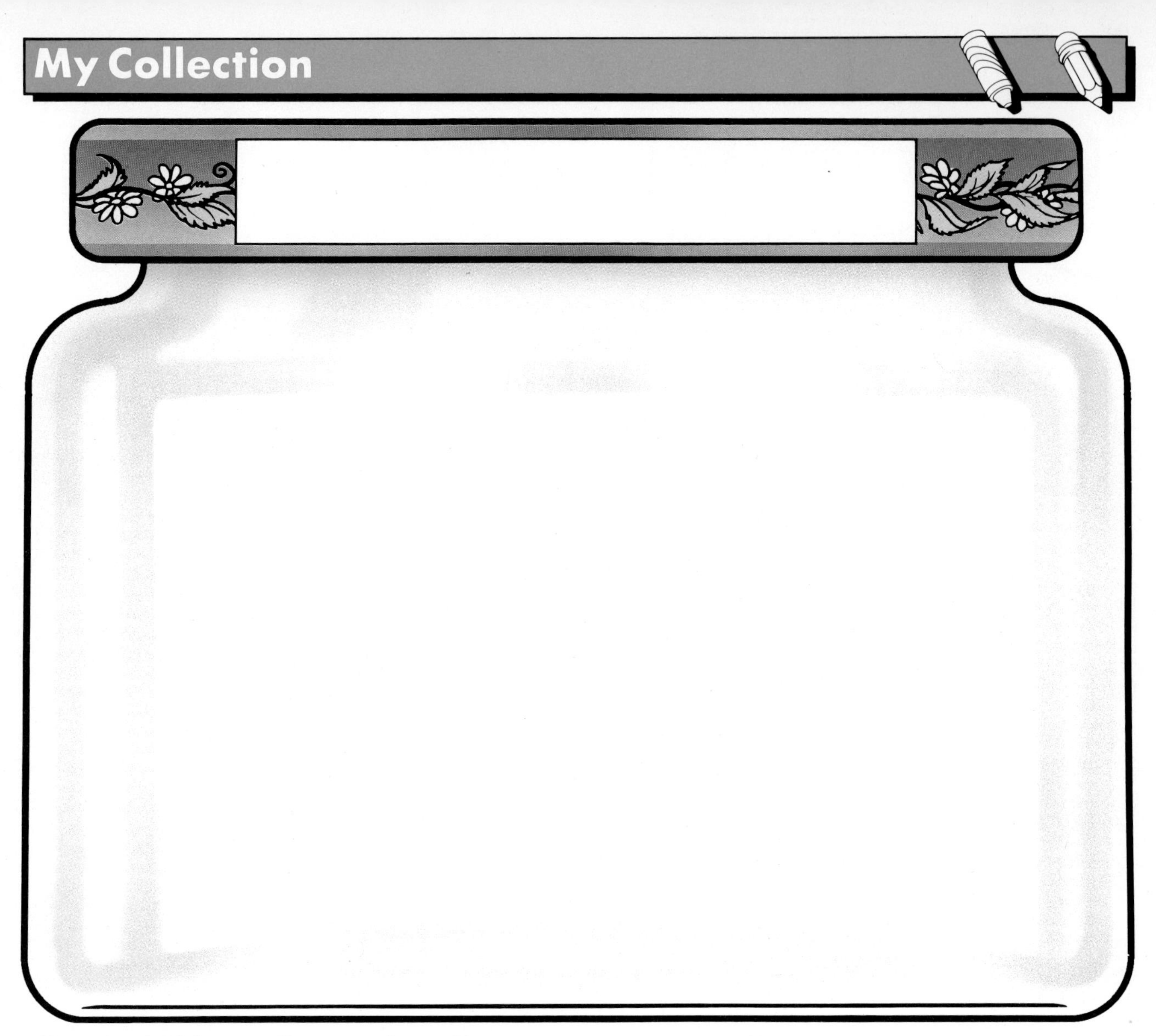

Have the child draw a collection and record how many items are in the collection.

Have the child draw a collection and record how many items are in the collection.

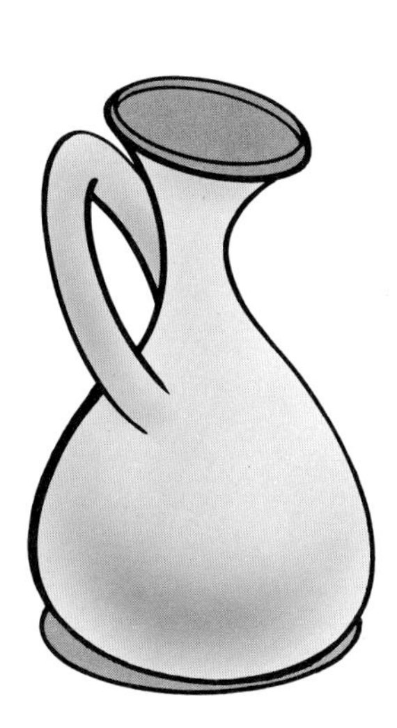

Ted picked 10 flowers.

Ted found ______________.

Have the child draw a set of 10 flowers in the vase. The child selects the number and kind of items to draw in the second container.

I found ______________________.

Have the child draw a set of objects and record how many there are.

The Imps at Play

has more than 8

has less than 7

has less than 9

has more than 6

Have the child draw a set of objects for each imp. Have the child tell how many he or she drew and why.

Imps in the Toy Chest

1 6 10

Have the child print the missing numerals and read them aloud.

Before

After

Have the child draw pictures to show what happened before and after Ted knocked on the door. Discuss the sequence of events with the child.

A Day with Ted

first

second

third

fourth

fifth

Have the child order the events in Ted's day by matching each word to a picture. Encourage the child to describe the sequence.

My Day

Have the child draw and label 3 events from his or her day in the order in which they happen. Encourage the child to discuss the sequence.

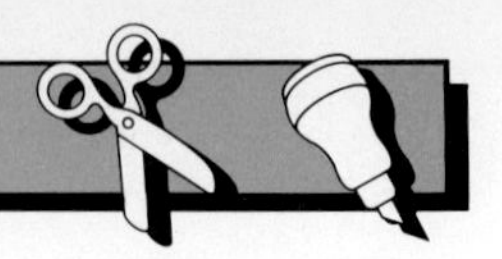

Have the child cut out the pieces of wool from page 31 and arrange them from shortest to longest. Discuss the differences in length.

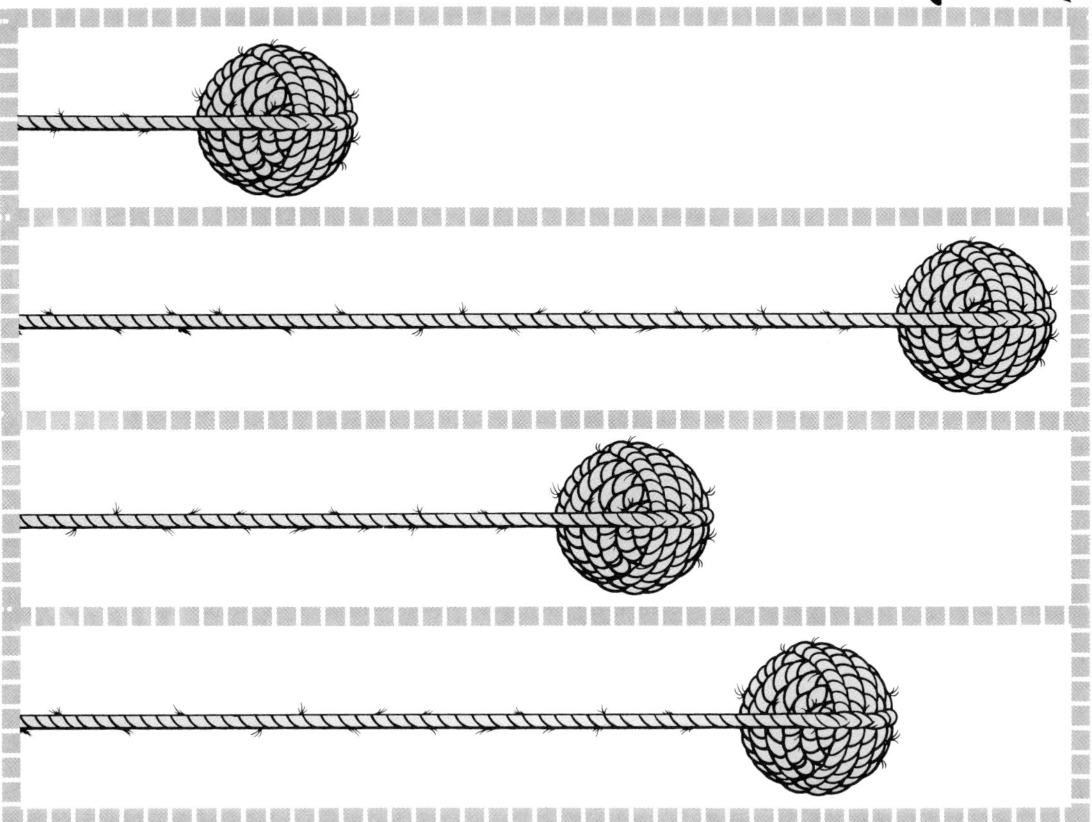

Have the child cut out the pieces of wool and order them by length on page 30.

Necklaces for You and Me

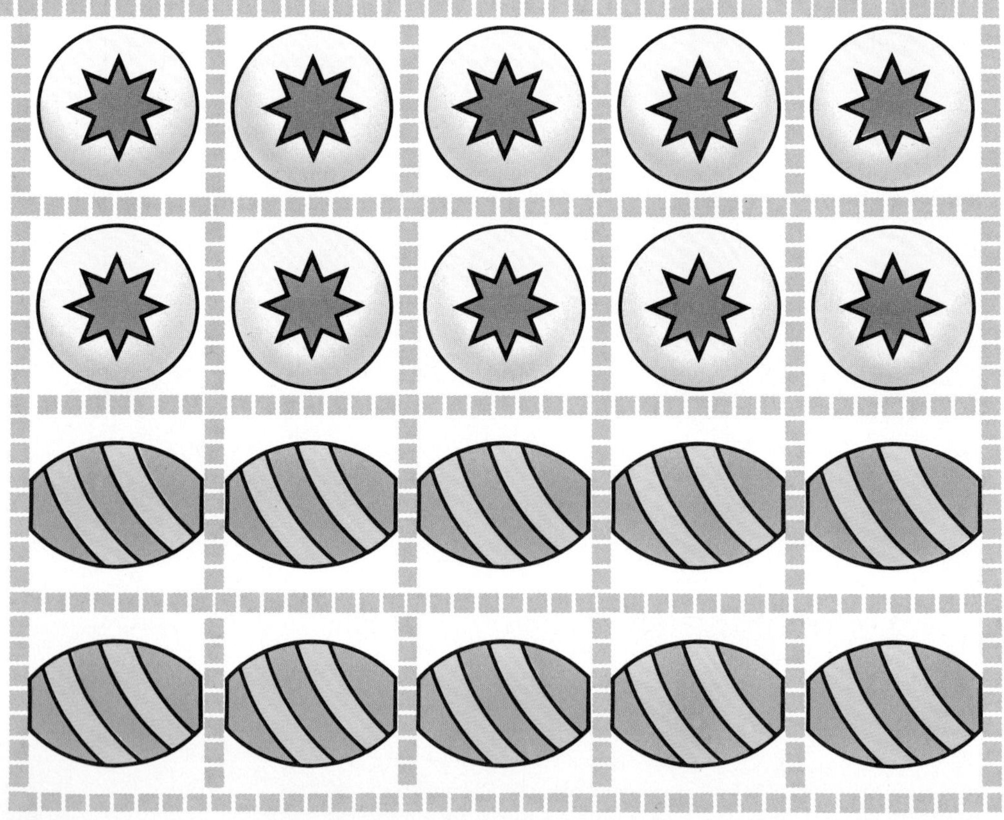

Have the child cut out the beads and arrange them in combinations of 4 on page 33.

Necklaces for You and Me

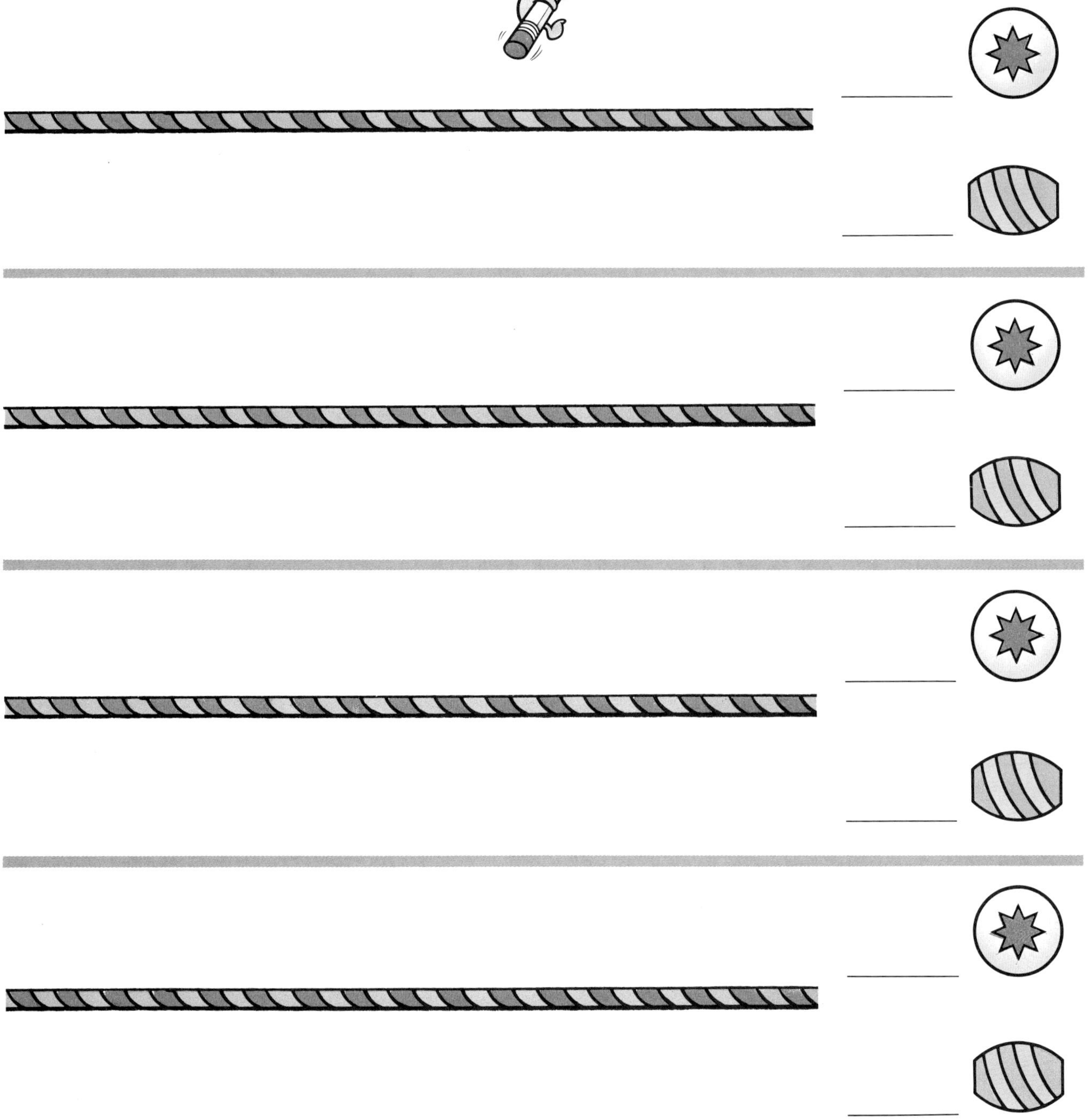

1 3 4

Have the child make and describe different combinations of 4 by arranging the beads from page 32 on the necklaces.

Floating Imps

Have the child draw and describe combinations of 5 balloons in 2 colors. Encourage the child to describe them, e.g., The imp has 5 balloons. 2 are yellow and 3 are green.

4-Trains

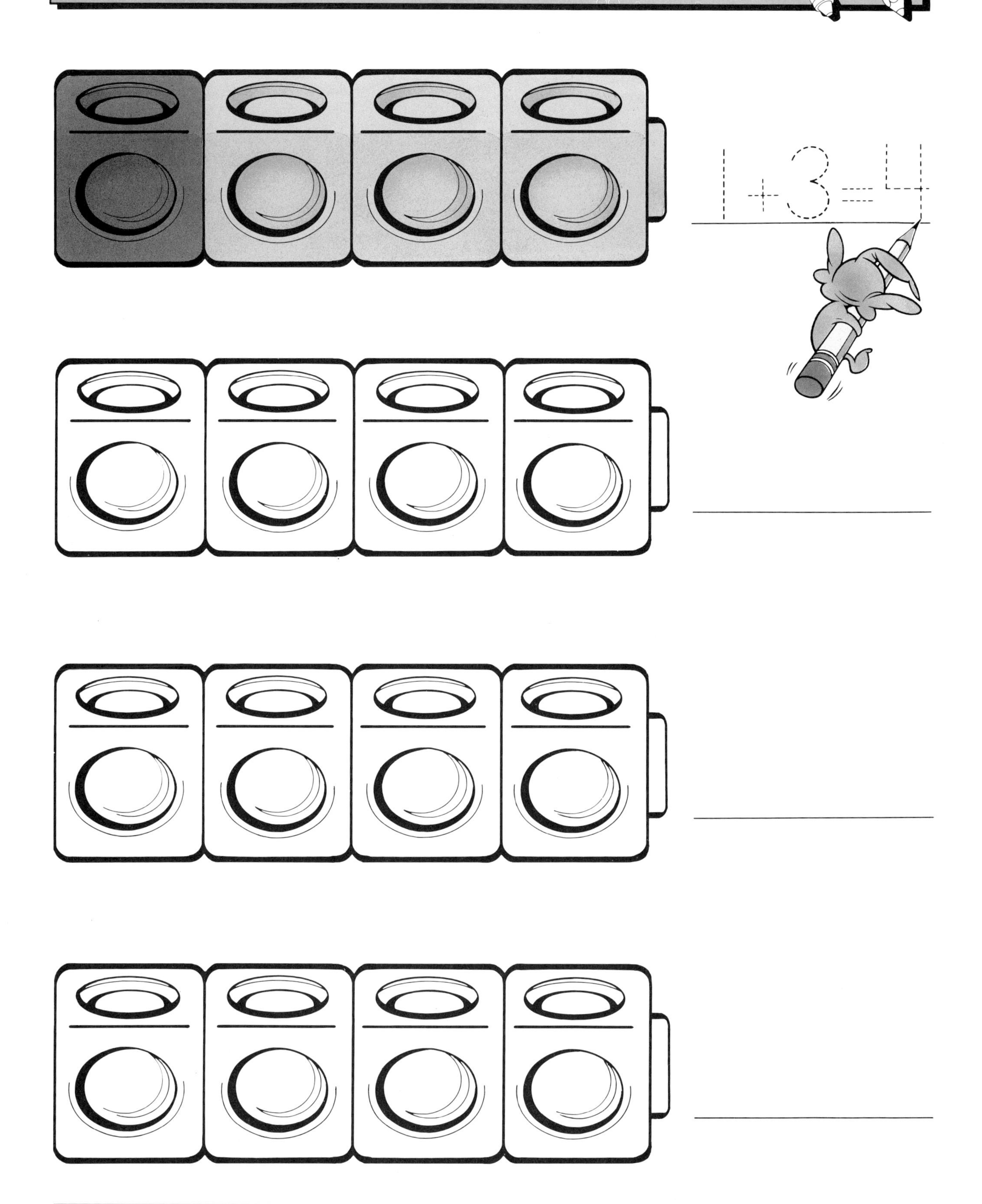

Have the child use 2 colors to make combinations of 4. Have the child print the number sentence that describes each 4-train.

5-Trains

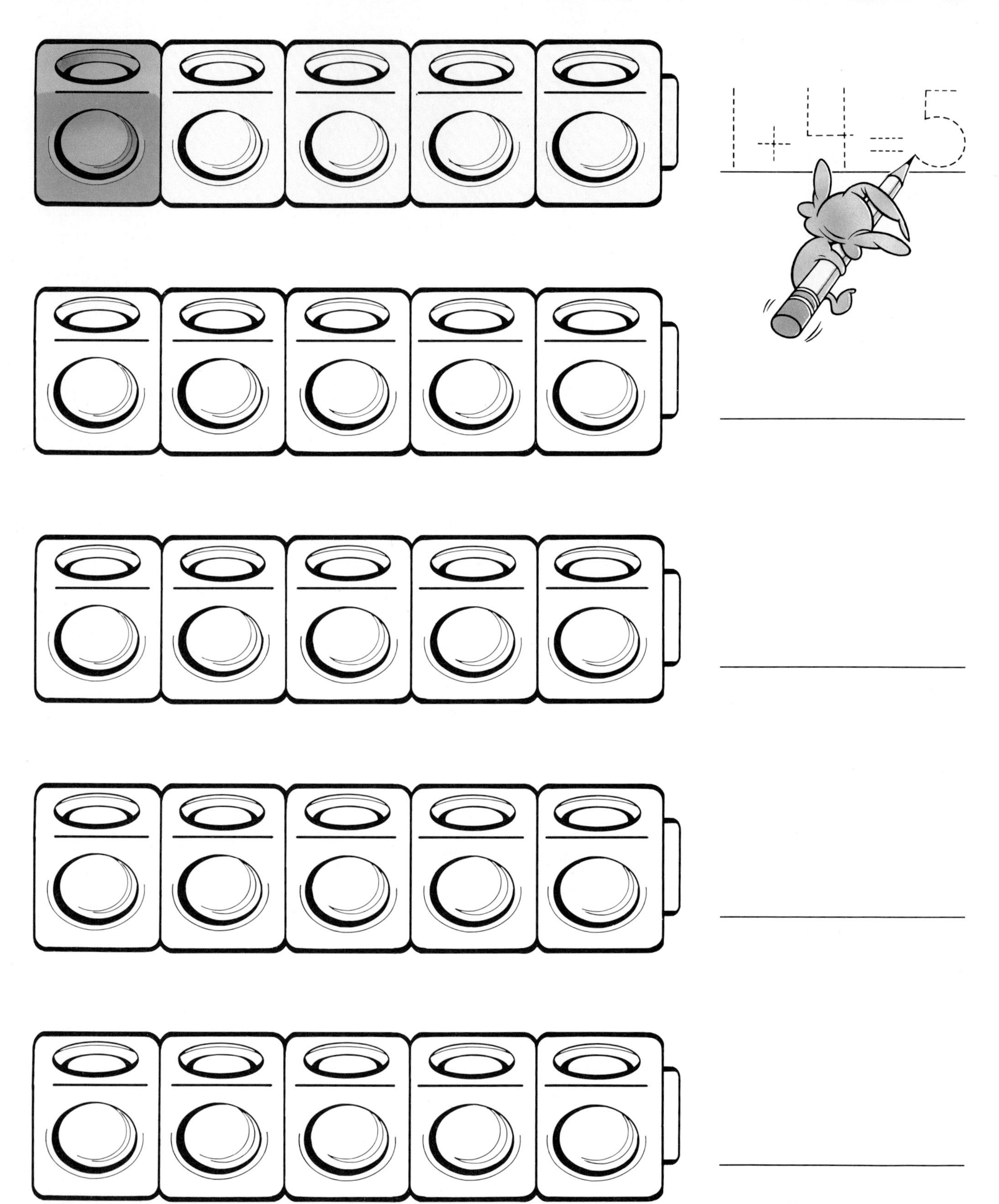

Have the child use 2 colors to make combinations of 5. Have the child print the number sentence that describes each 5-train.

6	7	8	9	10
16	17	18	19	20
26	27	28	29	30
36	37	38	39	40
46	47	48	49	50
56	57	58	59	60
66	67	68	69	70
76	77	78	79	80
86	87	88	89	90
96	97	98	99	100

Remove the hundreds chart (pages 37 and 56) from the book. Use as described on page 93.

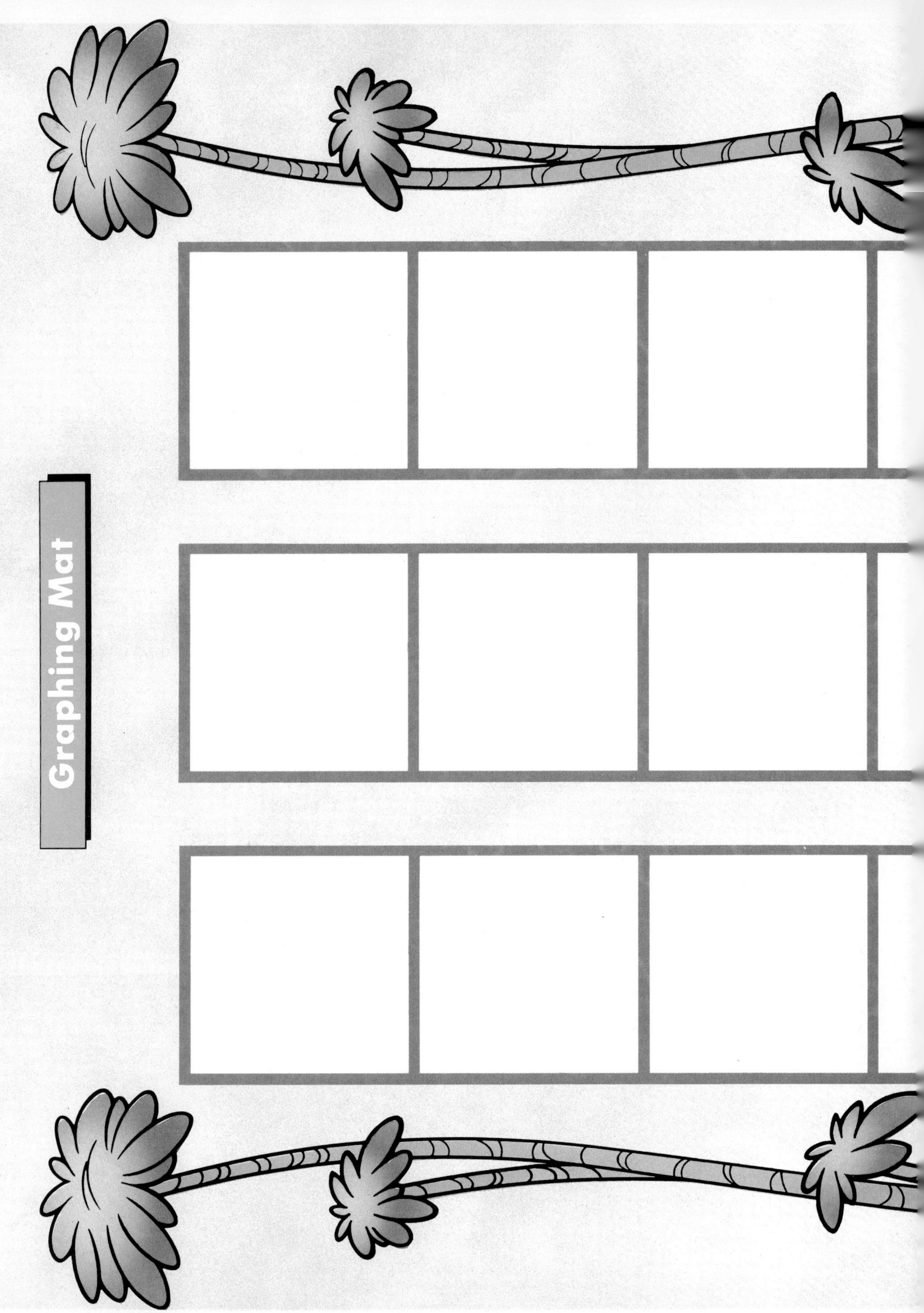

Remove the graphing mat (pages 38 and 55) from the book. Use as described on page 93.

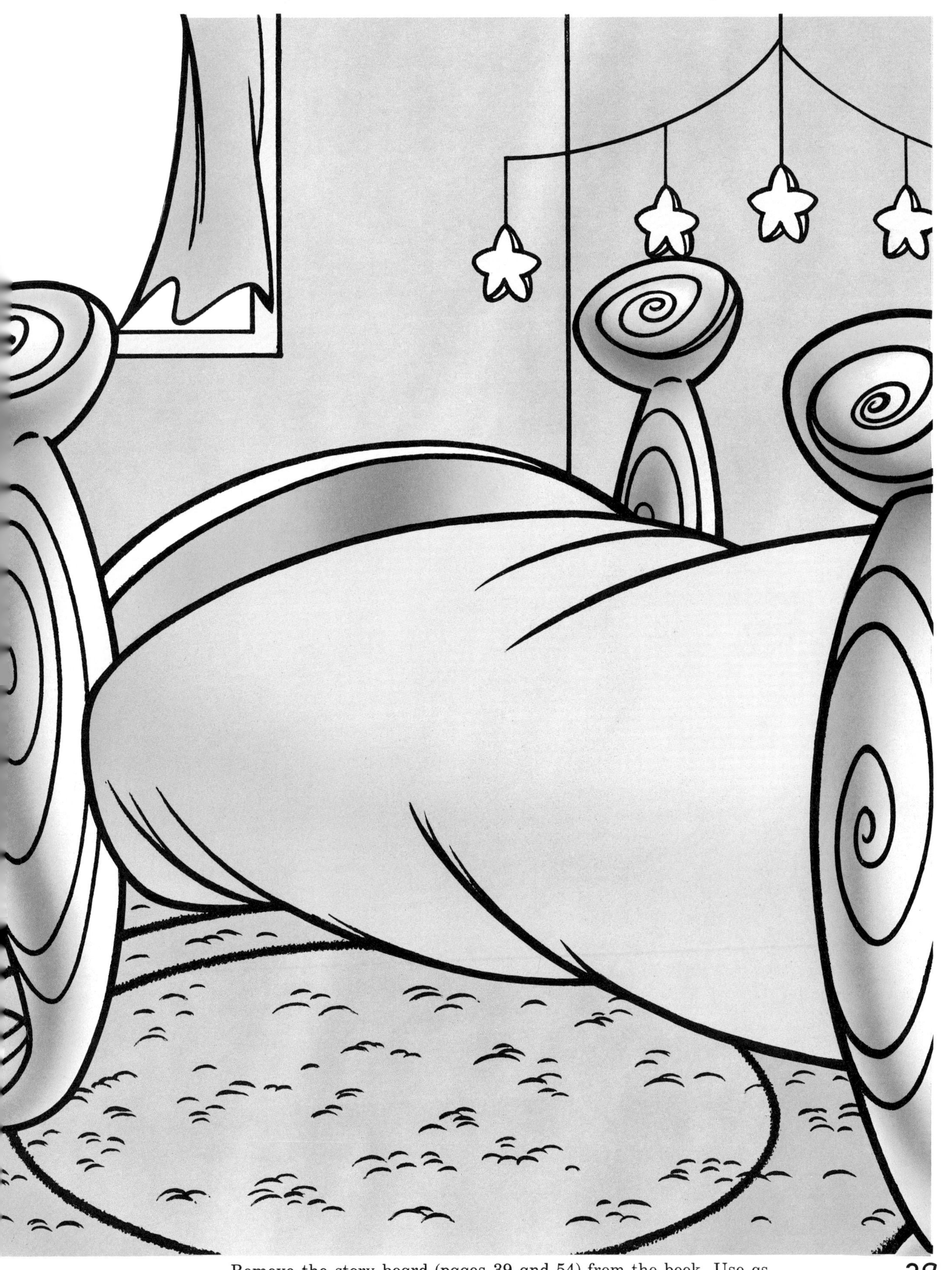

Remove the story board (pages 39 and 54) from the book. Use as described on page 93.

Remove the story board (pages 40 and 53) from the book. Use as described on page 93.

Remove the story board (pages 41 and 52) from the book. Use as described on page 93.

Remove the story board (pages 42 and 51) from the book. Use as described on page 93.

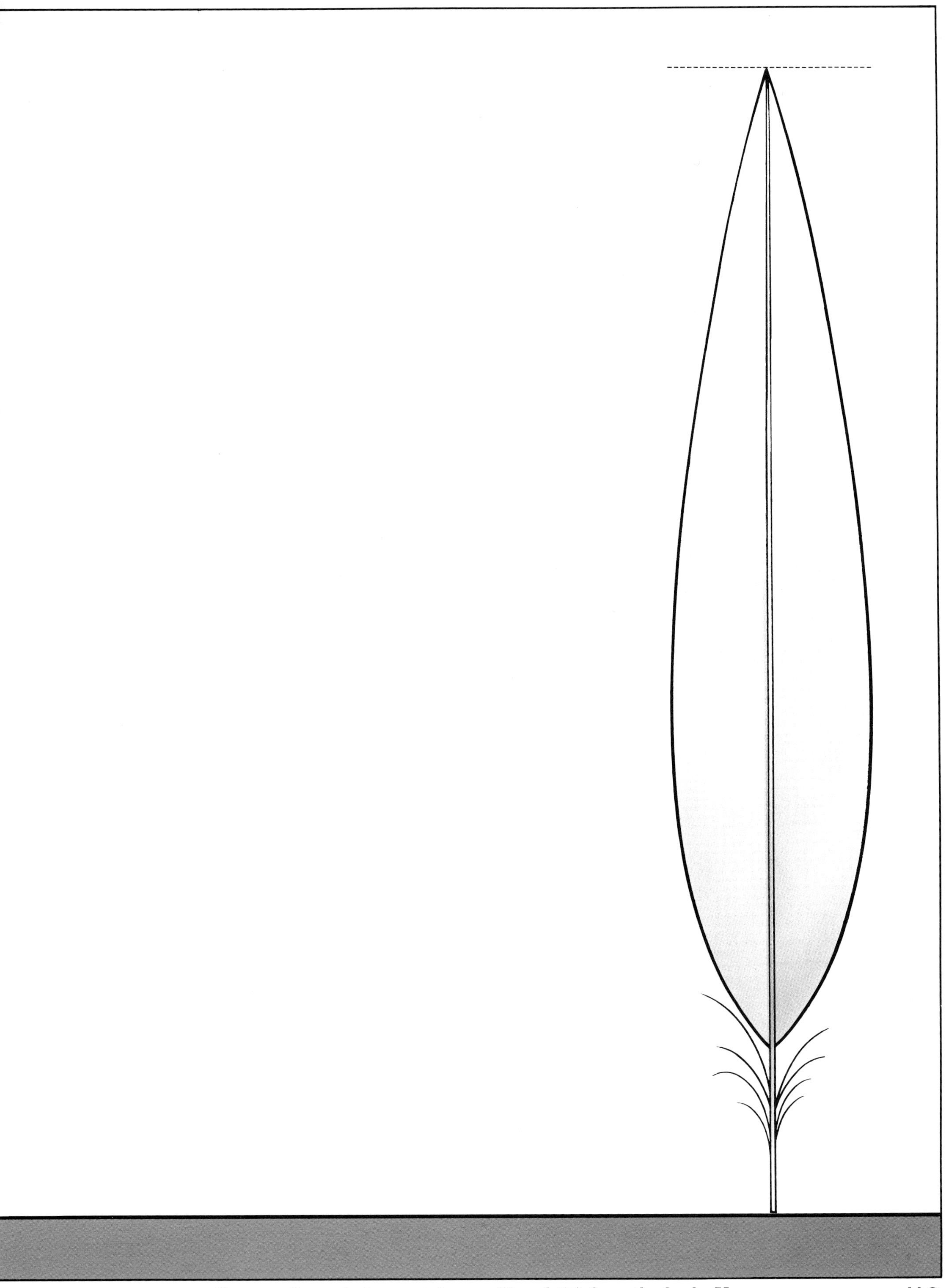

Remove the ordering mat (pages 43 and 50) from the book. Use as described on page 94.

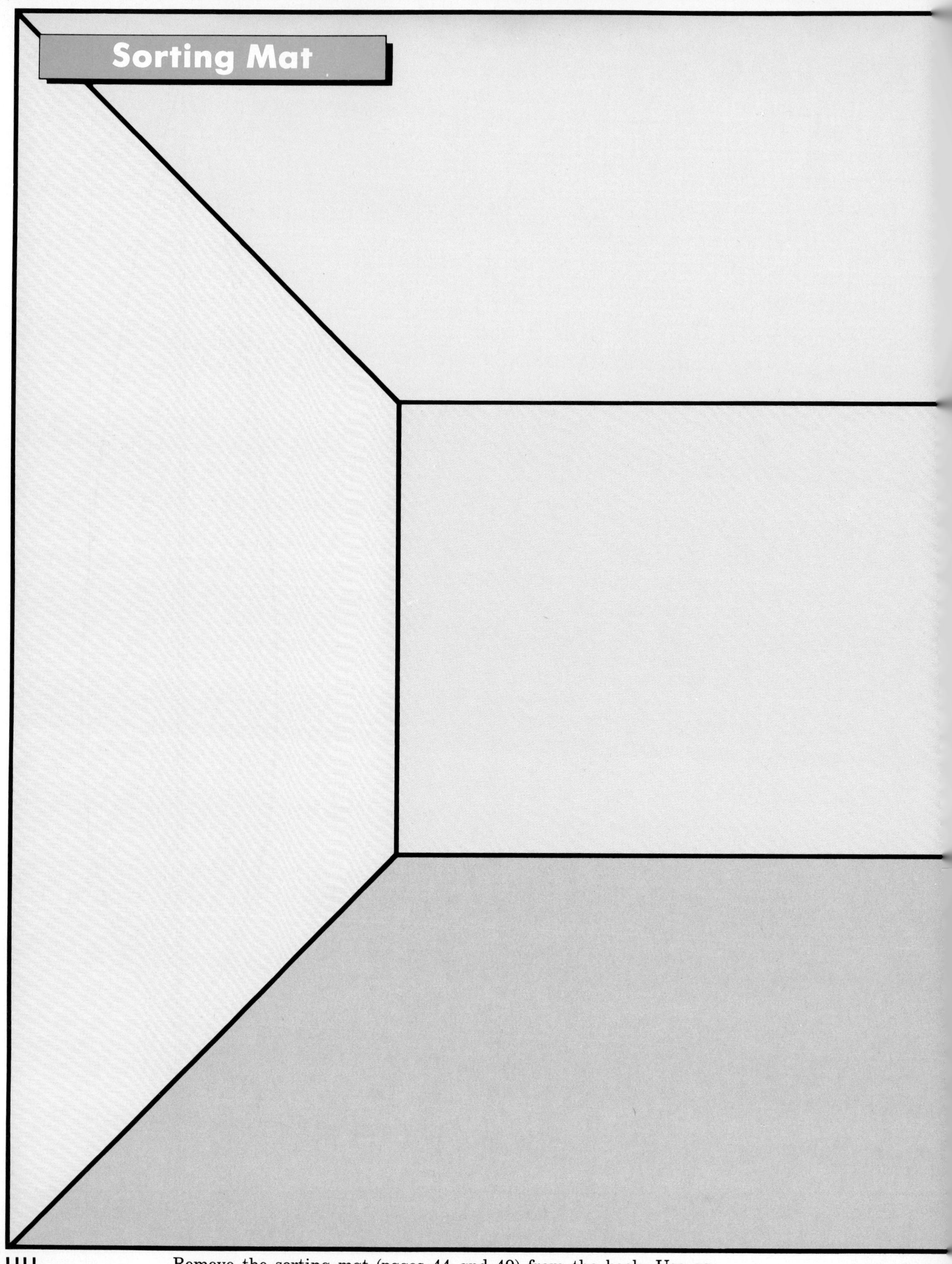

Remove the sorting mat (pages 44 and 49) from the book. Use as described on page 94.

Remove the game board (pages 45 and 48) from the book. Use as described on page 94.

Remove the game board (pages 46 and 47) from the book. Use as described on page 94.

1
2
3
4
5

See page 45 for directions.

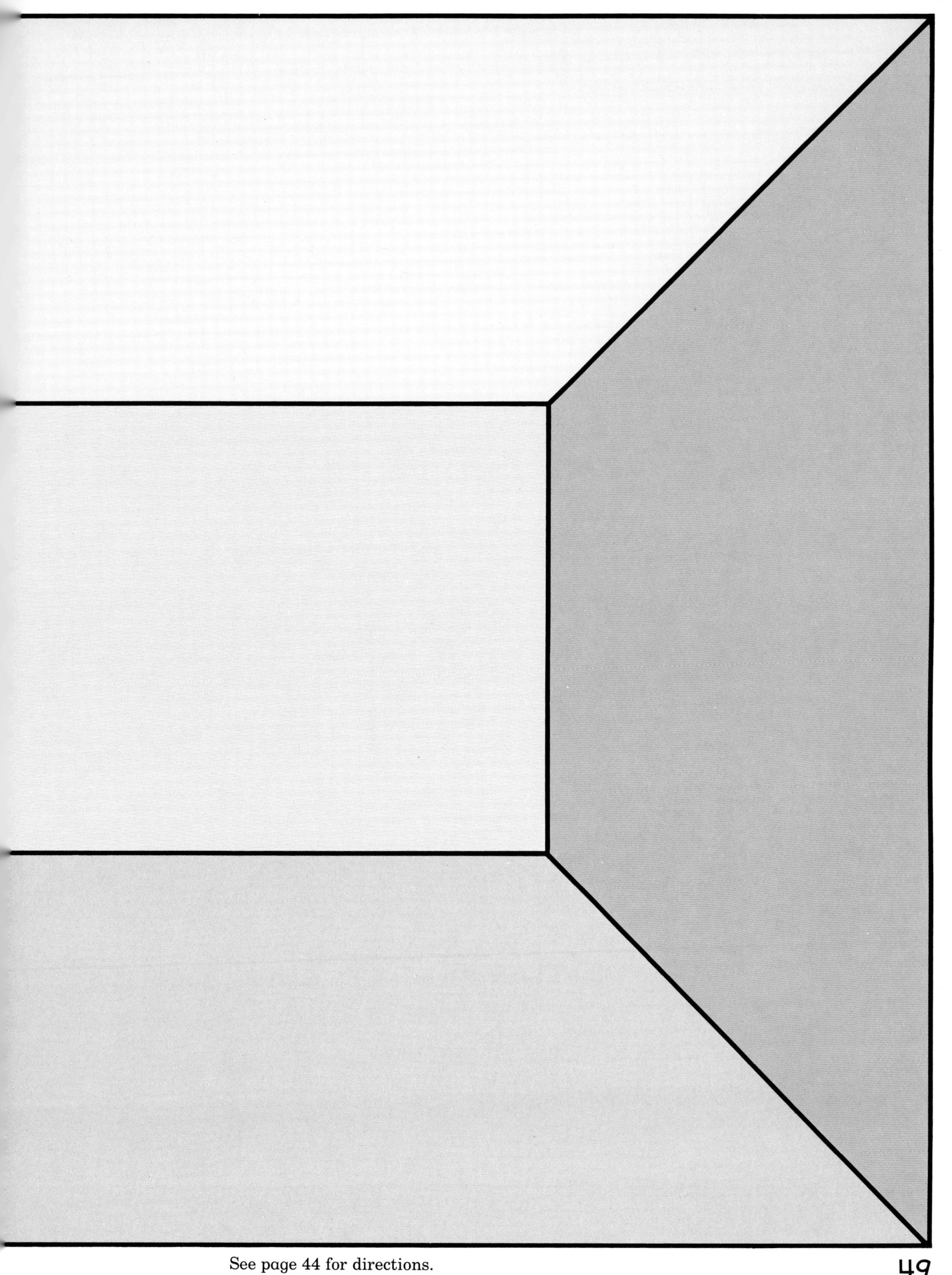

See page 44 for directions.

Ordering Mat

See page 43 for directions.

See page 42 for directions.

The Meadow

See page 41 for directions.

See page 40 for directions.

See page 39 for directions.

See page 38 for directions.

Hundreds Chart

1	2	3	4
11	12	13	14
21	22	23	24
31	32	33	34
41	42	43	44
51	52	53	54
61	62	63	64
71	72	73	74
81	82	83	84
91	92	93	94

See page 37 for directions.

The Imps Took One!

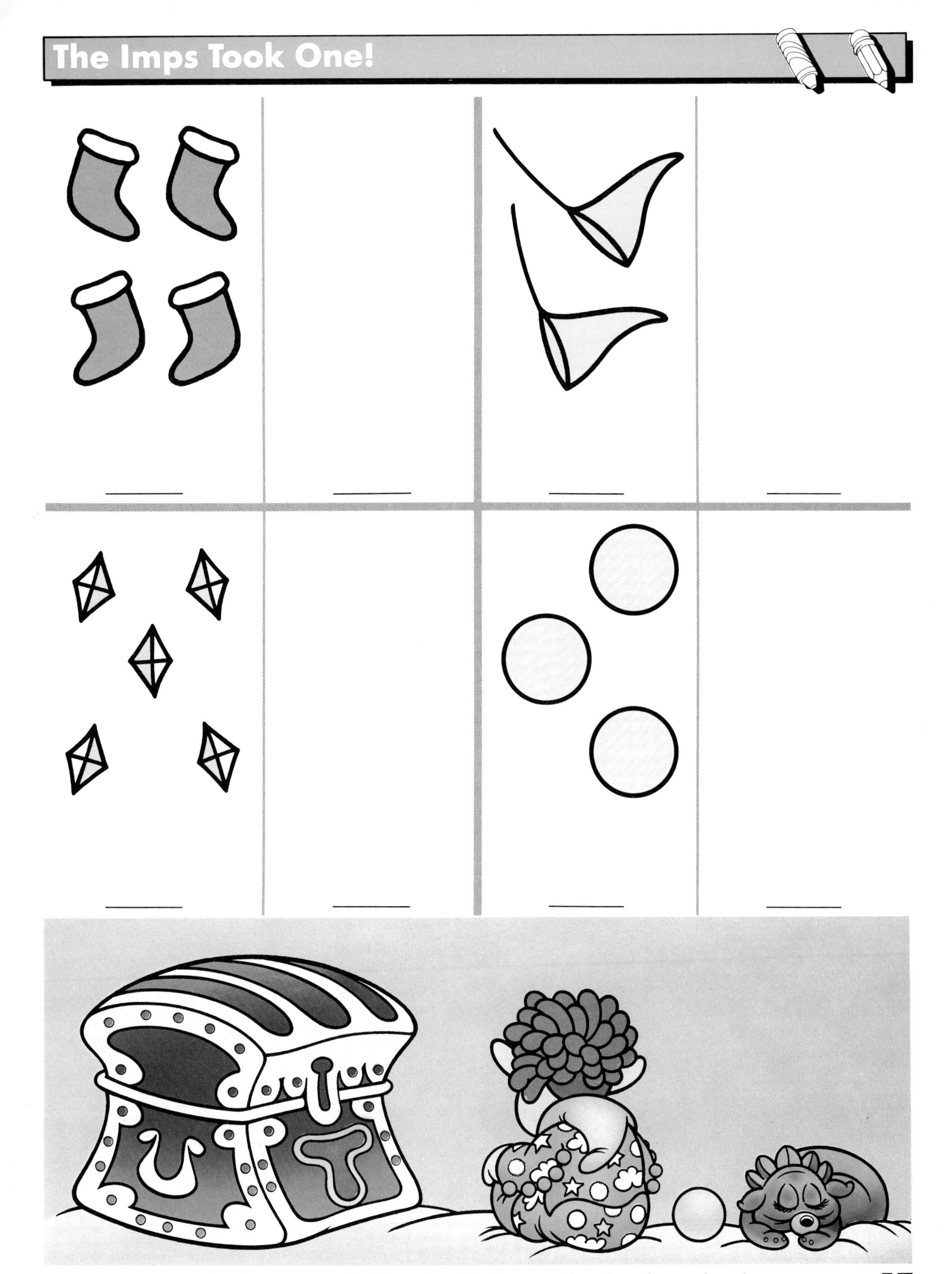

In the space provided, have the child draw one less than the number of items shown in each set, and record how many items are left.

Broken Balloons

4 balloons
1 was broken.
How many are left? ____

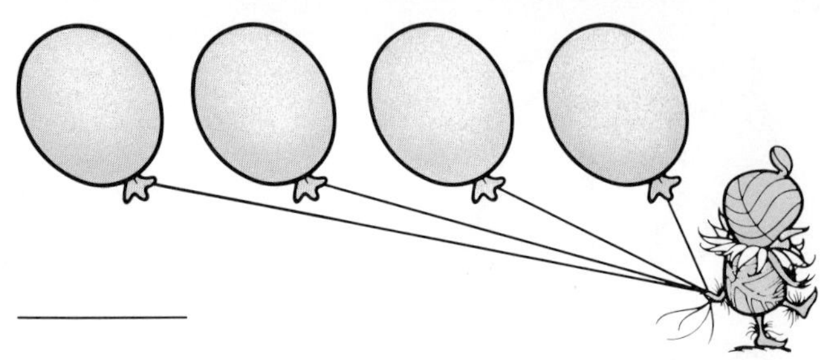

4 balloons
3 were broken.
How many are left? ____

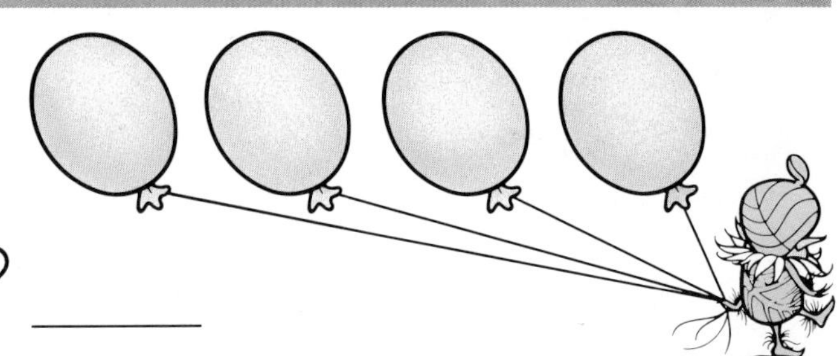

4 balloons
0 were broken.
How many are left? ____

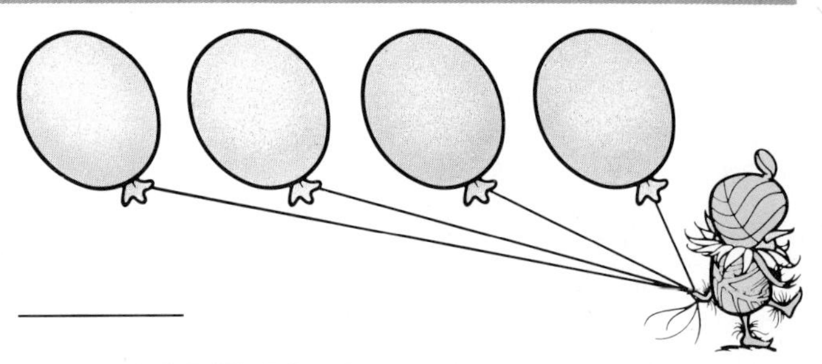

4 balloons
2 were broken.
How many are left? ____

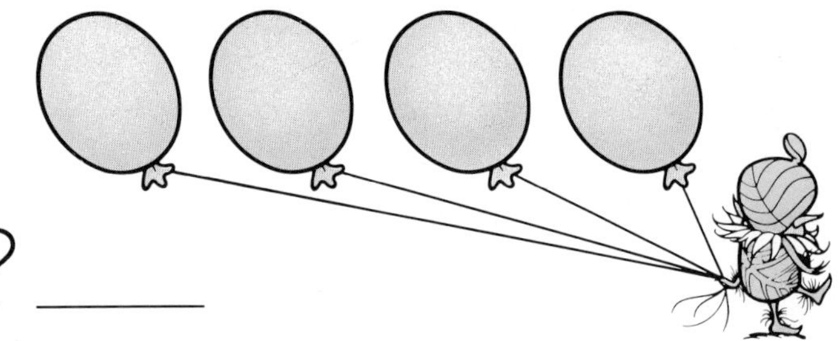

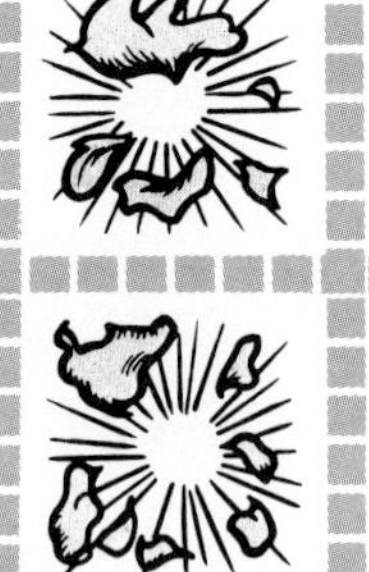
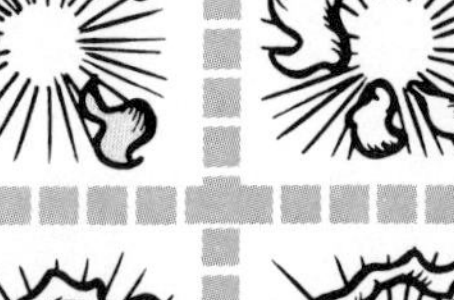
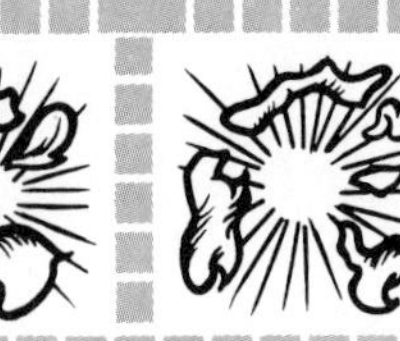

Have the child cut and glue broken balloons to cover the number of balloons indicated, then record how many are left.

Midnight Snack

5 cookies

3 were eaten.

How many are left? ____

5 muffins

2 were eaten.

How many are left? ____

5 toast

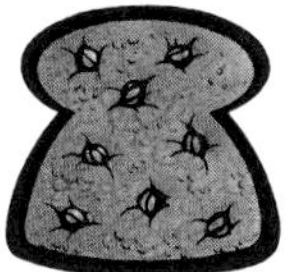    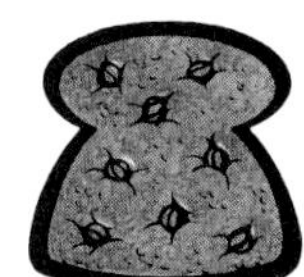

5 were eaten.

How many are left? ____

5 crackers

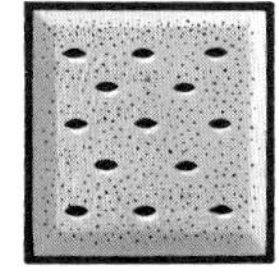 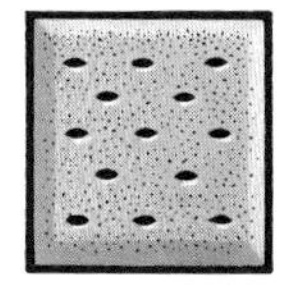 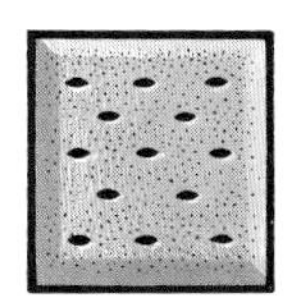 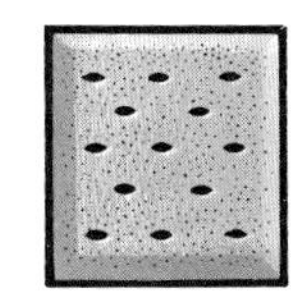 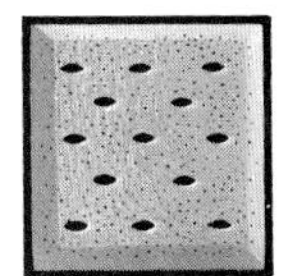

4 were eaten.

How many are left? ____

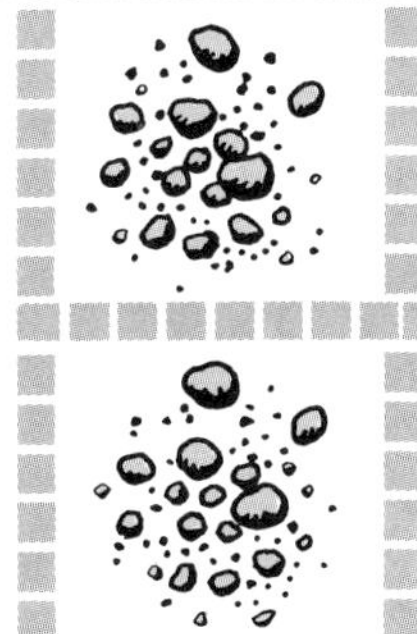 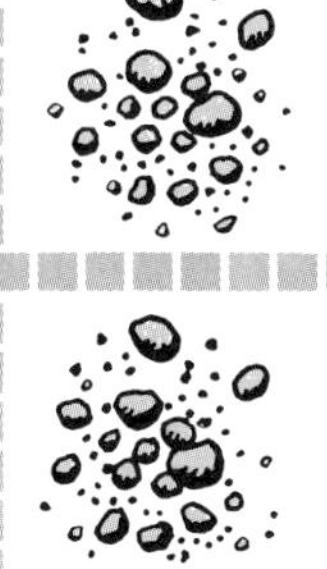 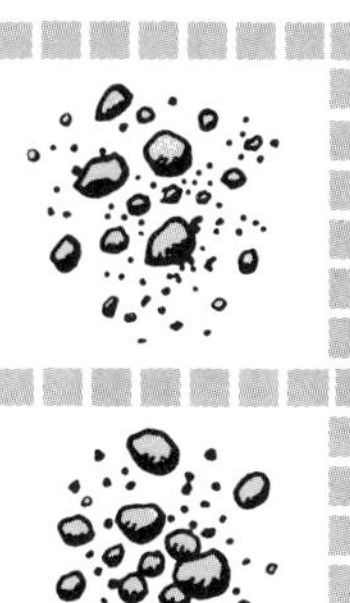 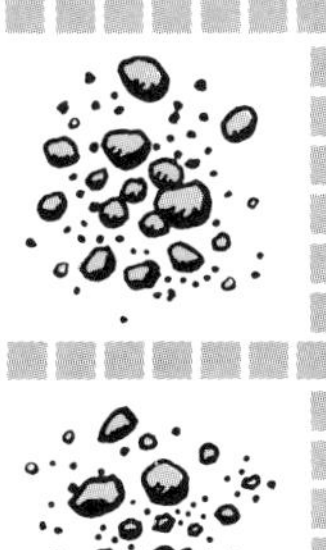 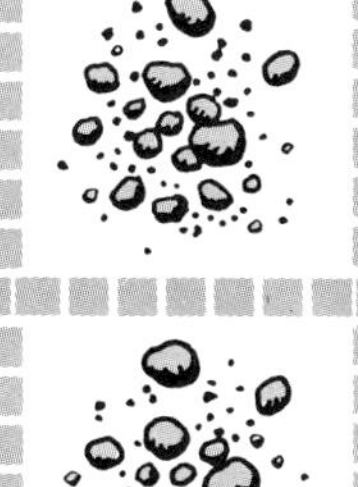 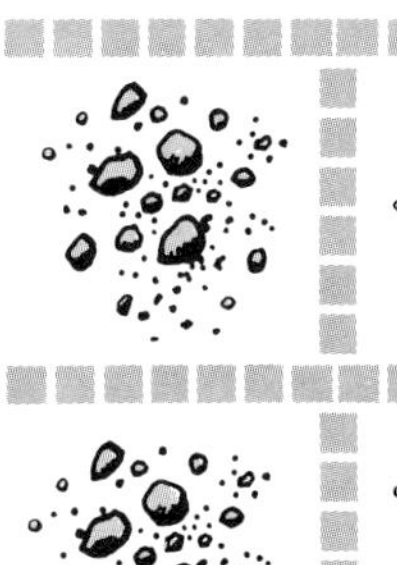 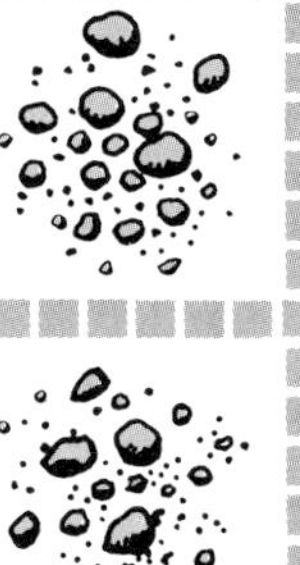

Have the child glue crumbs to cover the foods eaten, then record the number left.

Hiding Ted's Toys

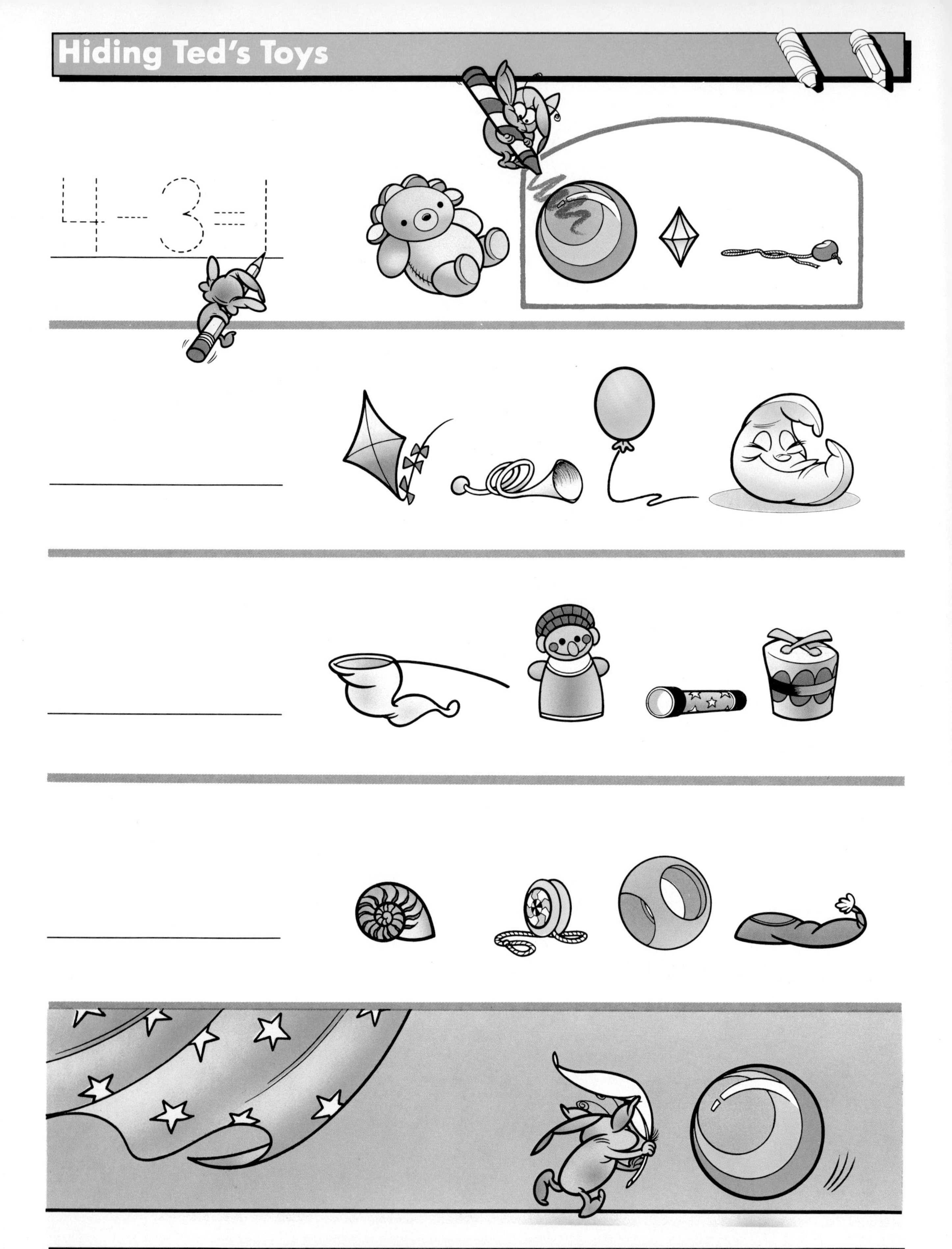

Have the child hide some of Ted's toys by drawing a box around them, then print the number sentence.

Hide the Imps!

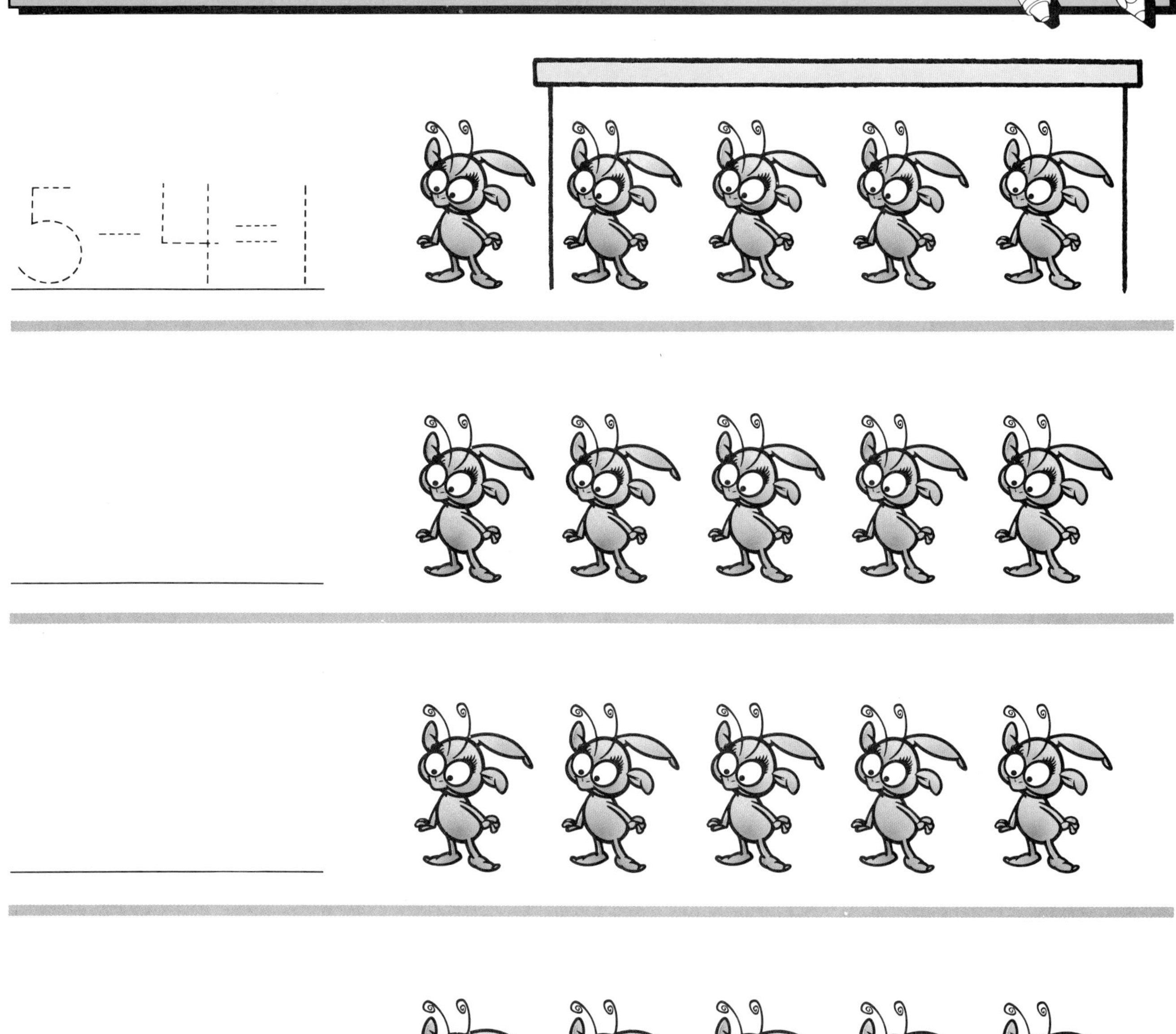

Have the child hide the imps by drawing a shelter over them, then print the number sentence.

Kite Flying

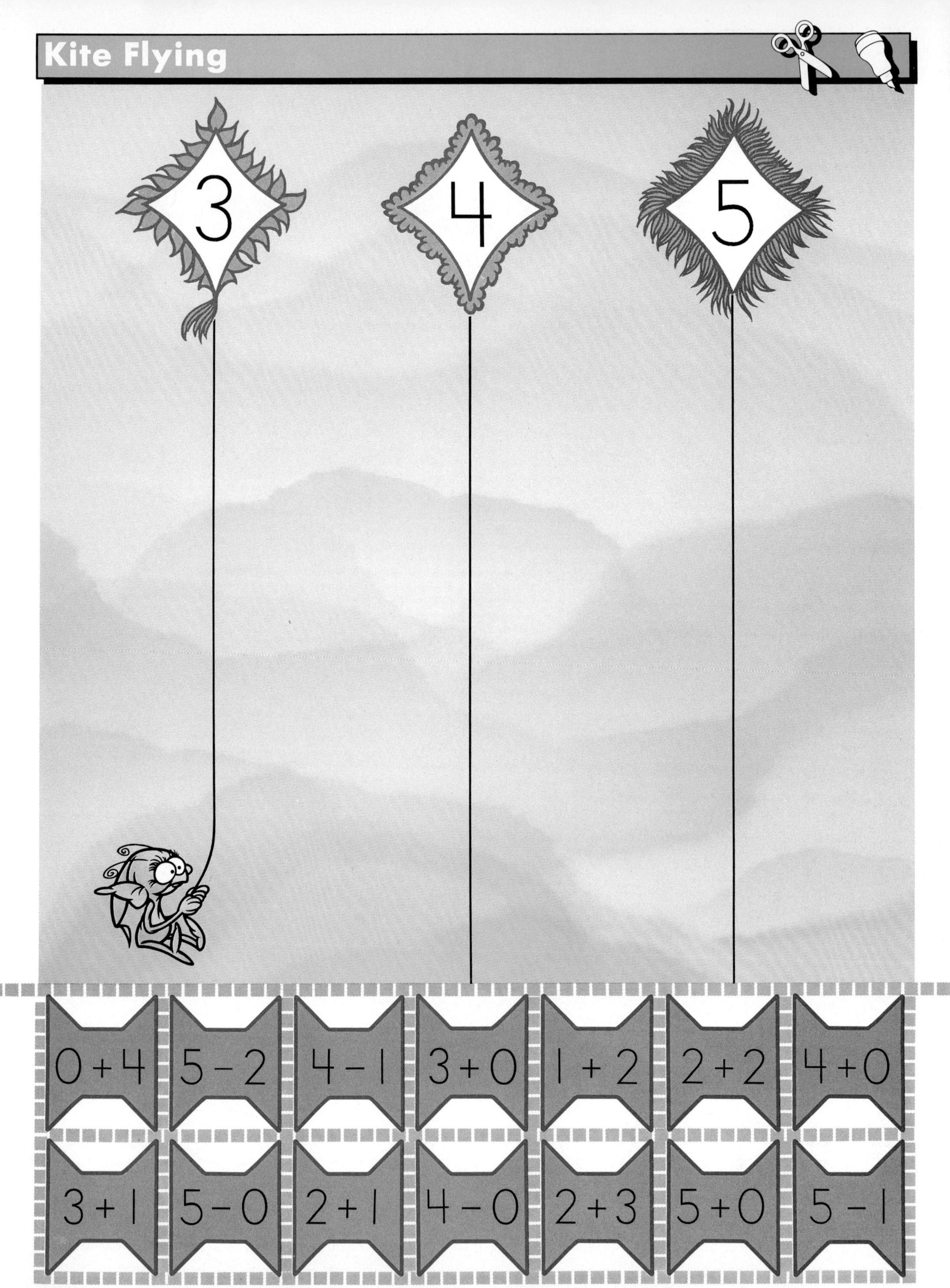

Have the child cut out the number expressions and place each on the appropriate kite.

Look Around You!

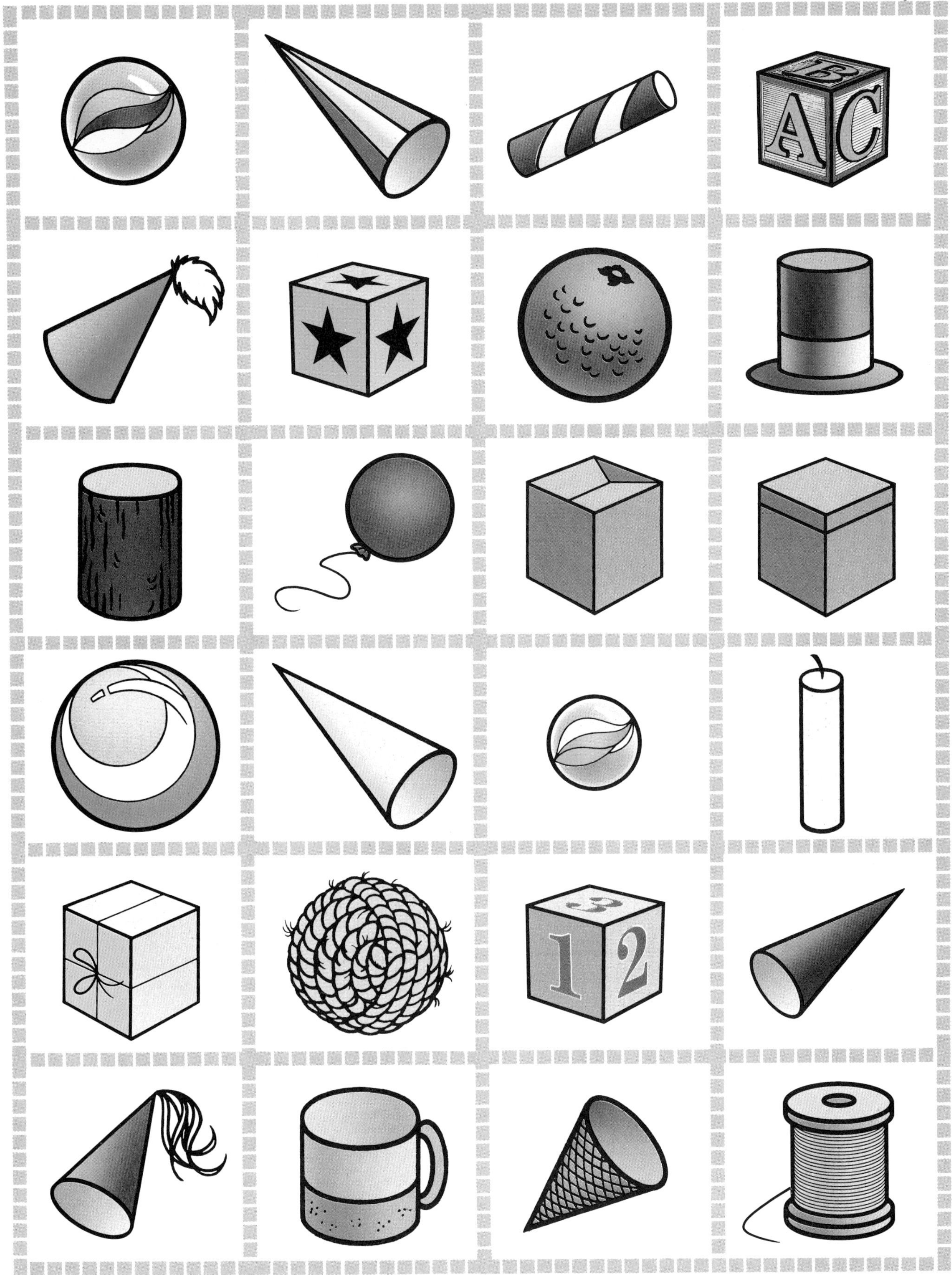

Have the child cut out the objects and sort them by shape onto pages 65 and 66.

Look Around You!

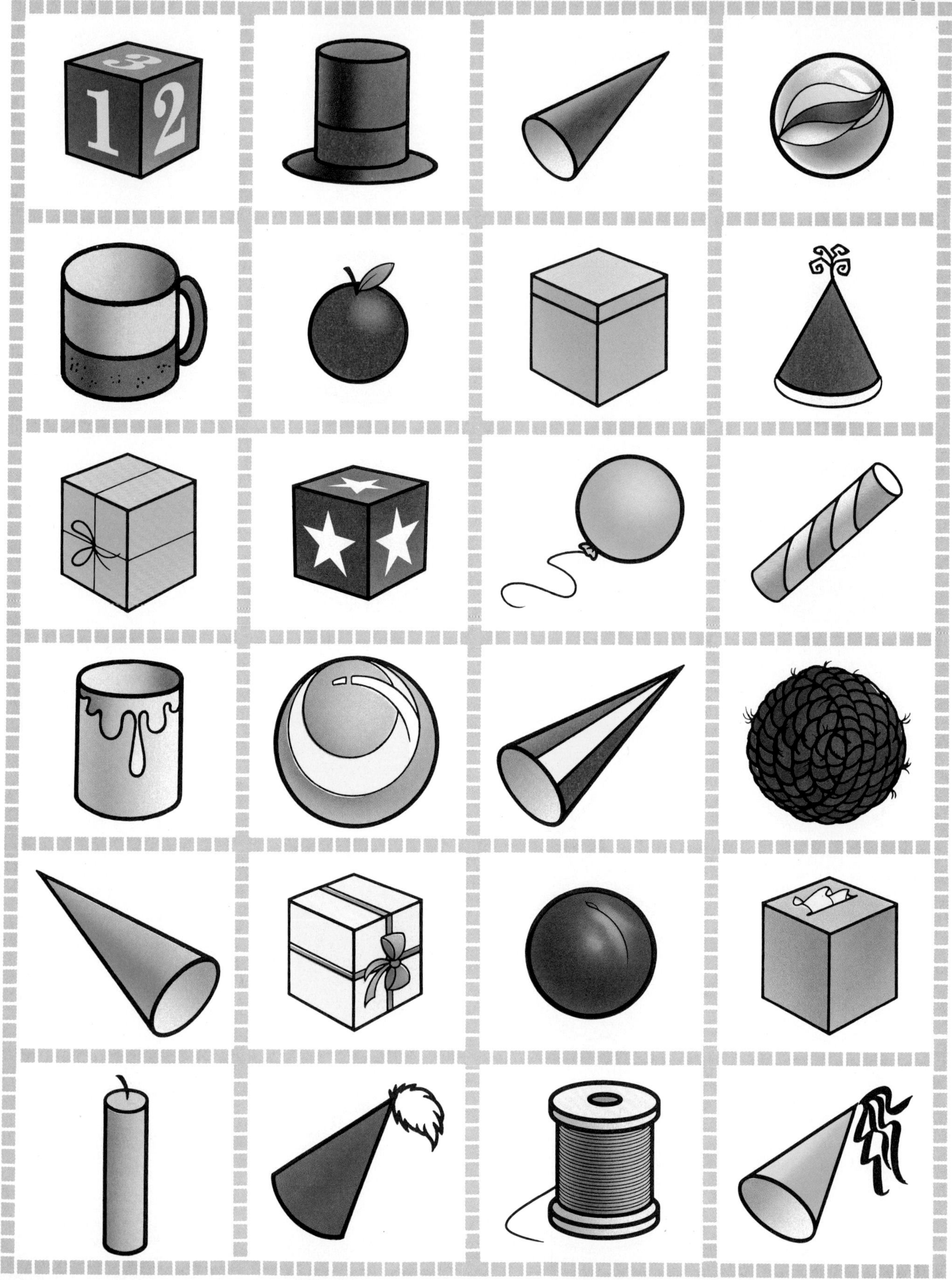

Have the child cut out the objects and sort them by shape onto pages 65 and 66.

Look Around You!

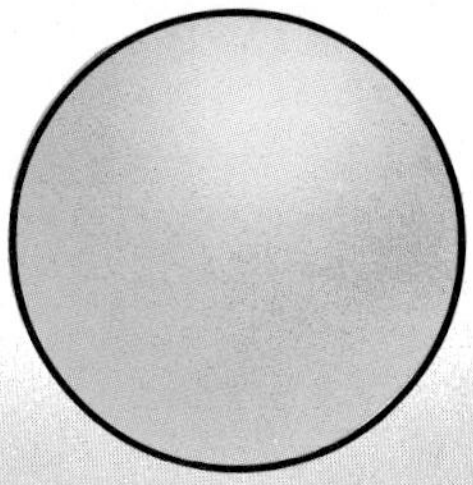 sphere	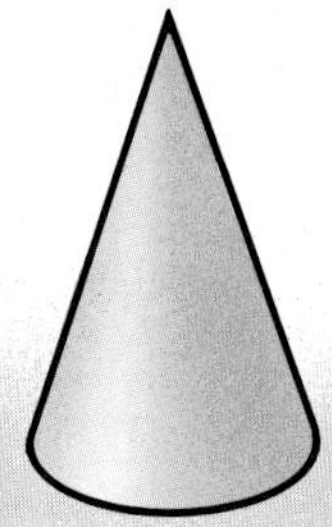 cone

Have the child sort and place the objects from pages 63 and 64 in the appropriate column on pages 65 and 66.

Look Around You!

cube	cylinder

Have the child sort and place the objects from pages 63 and 64 in the appropriate column on pages 65 and 66.

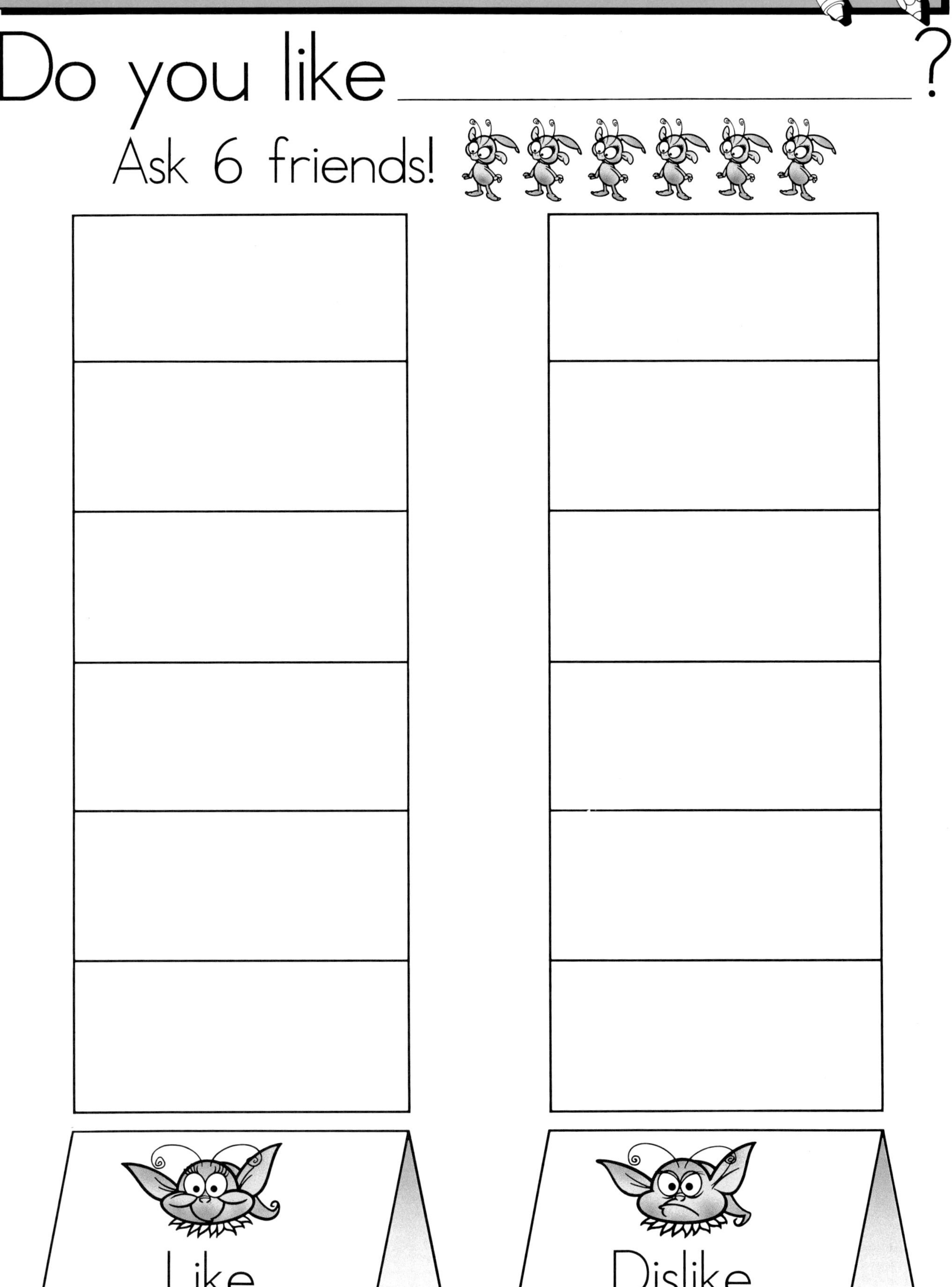
Do you like ____________________?
Ask 6 friends!
Like
Dislike

Have the child complete the question and present it to 6 friends. A happy or sad face is drawn in the appropriate column in response. Discuss the results of the graph with the child.

Which Do You Like?

Do you like ________________________

______________ or ______________ ?

Ask some friends!

I asked ______________ friends.

Most like ______________.

Have the child complete the question and labels and present the question to some friends. A picture is drawn in the appropriate columns in response. Discuss the results of the graph with the child.

Game Pieces

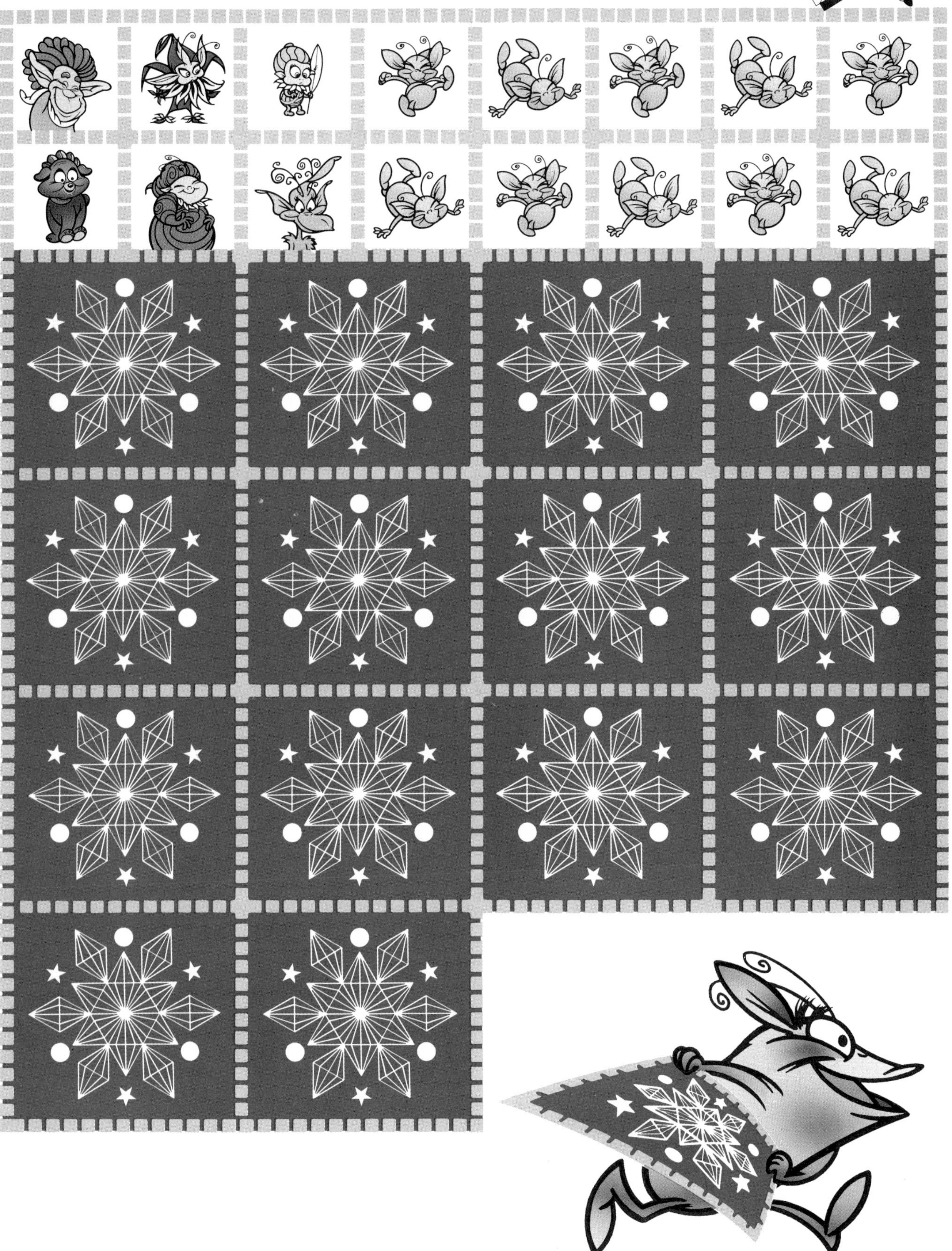

Have the child cut out the game pieces and store them in an envelope.
See page 94 for further directions.

Game Pieces

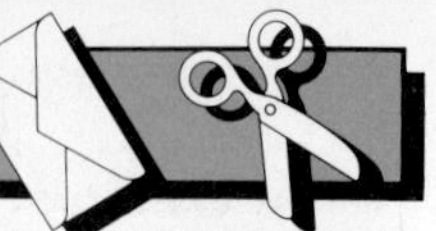

See directions on the preceding page.

Bouncing on the Bed!

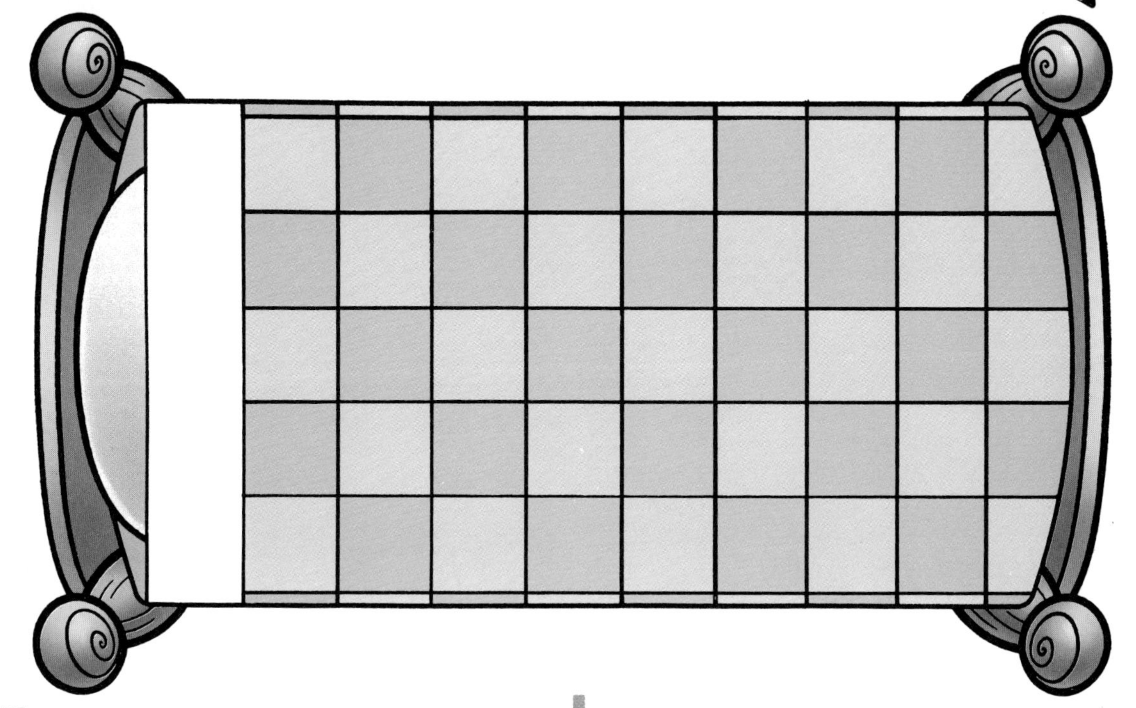

5 Imps bouncing!

Front	Back	
		Imps on the bed.

3 + 2 = 5

___ + ___ = ___

___ + ___ = ___

___ + ___ = ___

___ + ___ = ___

___ Imps bouncing!

Front	Back	
		Imps on the bed.

___ ___ ___

___ ___ ___

___ ___ ___

___ ___ ___

___ ___ ___

Have the child drop 5 of the imps from pages 69 and 70 onto the bed to create and record facts of 5, then repeat for another number of his or her choice.

Imps at Play

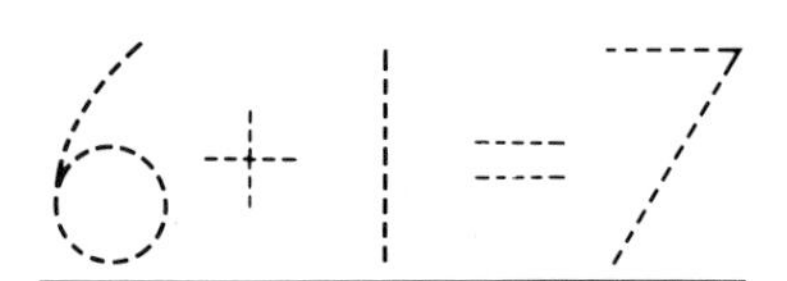

Have the child describe each story and record the corresponding number sentence.

Imps at Play

How will you end this story?

Use as described on the preceding page. Discuss how the story might end and have the child draw the final scene and label it with a number sentence.

Candles Burning

9 - 6 =

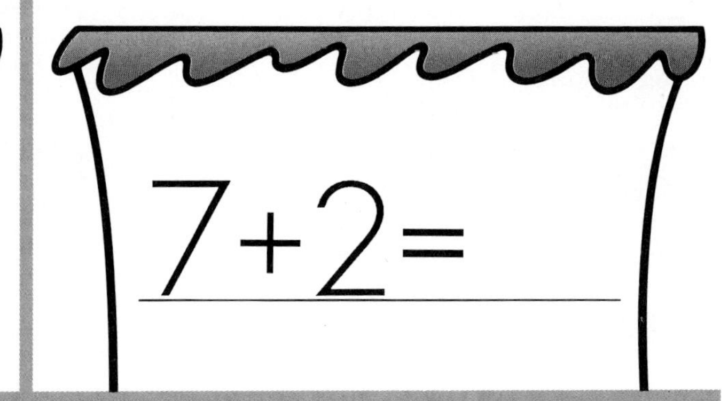

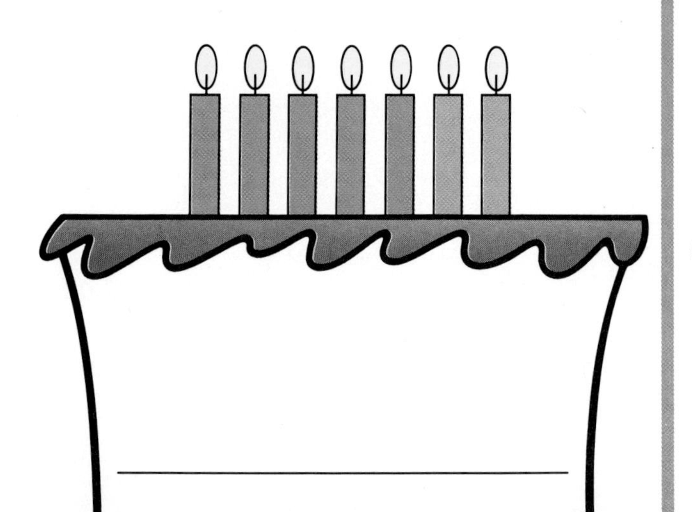

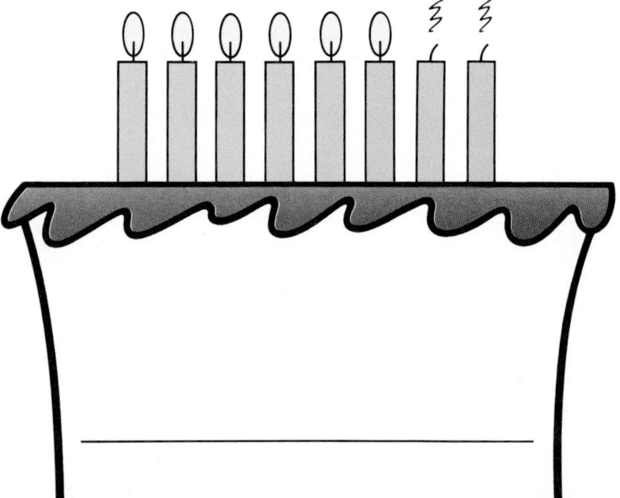

Have the child draw the candles to match the number sentence or print the sentence to match the candles.

Candles Burning

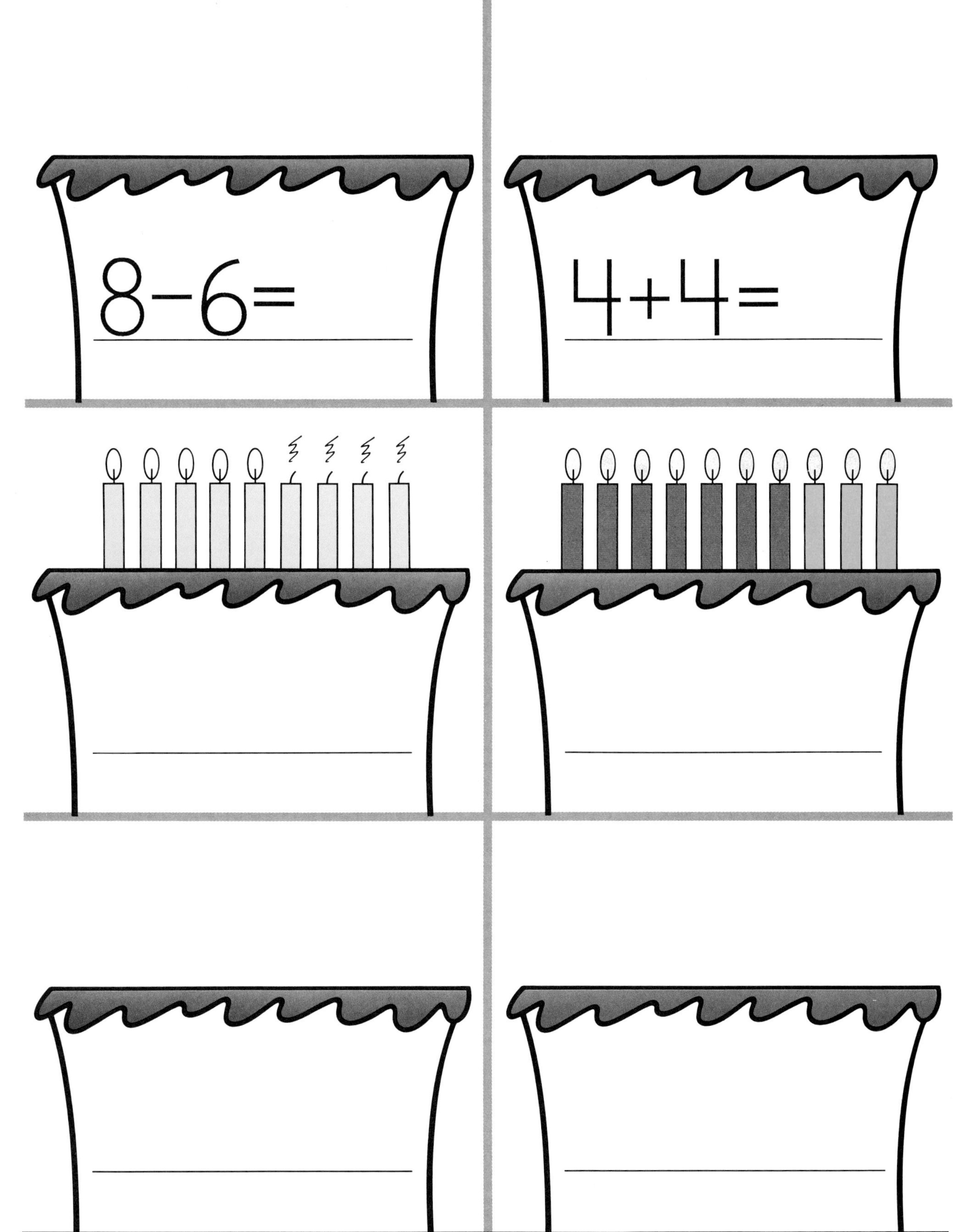

Use as described on the preceding page. Encourage the child to create her or his own candle stories for the last two cakes and print the corresponding number sentence.

A Sunny Day

Have the child cut out the figures from pages 77 and 78 and arrange them onto the picture on pages 76 and 79. Encourage the child to record the number of each kind of figure used to complete the scene.

A Sunny Day

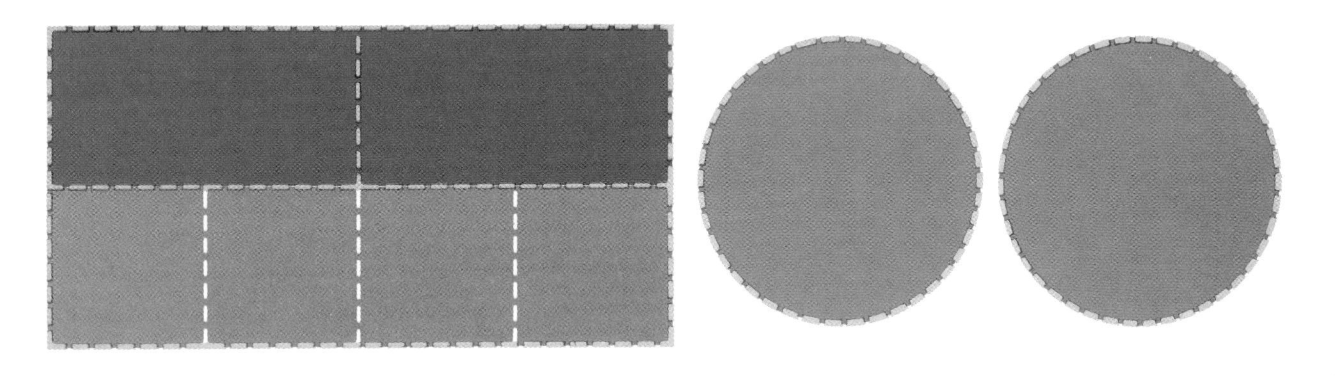

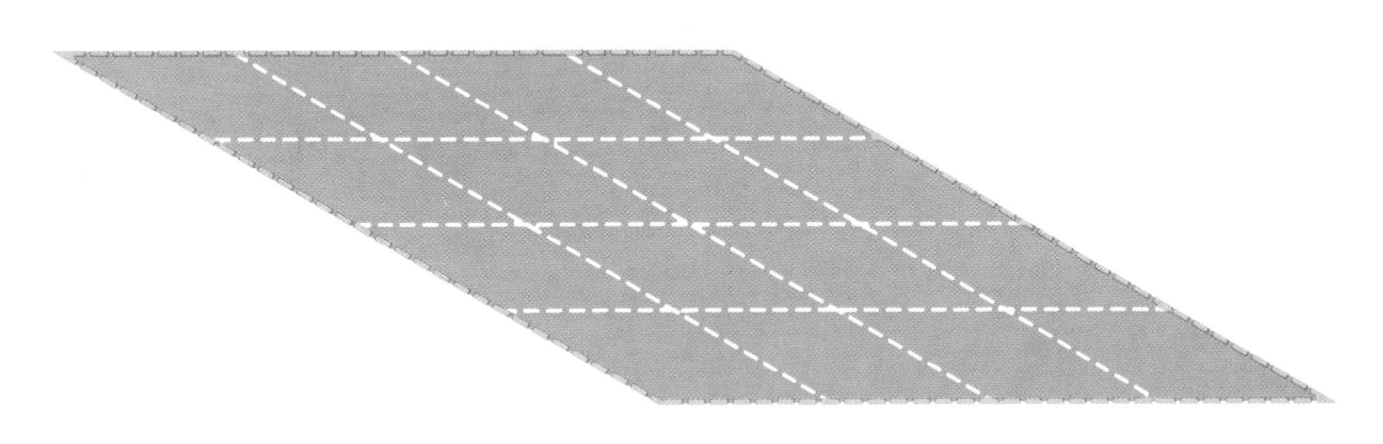

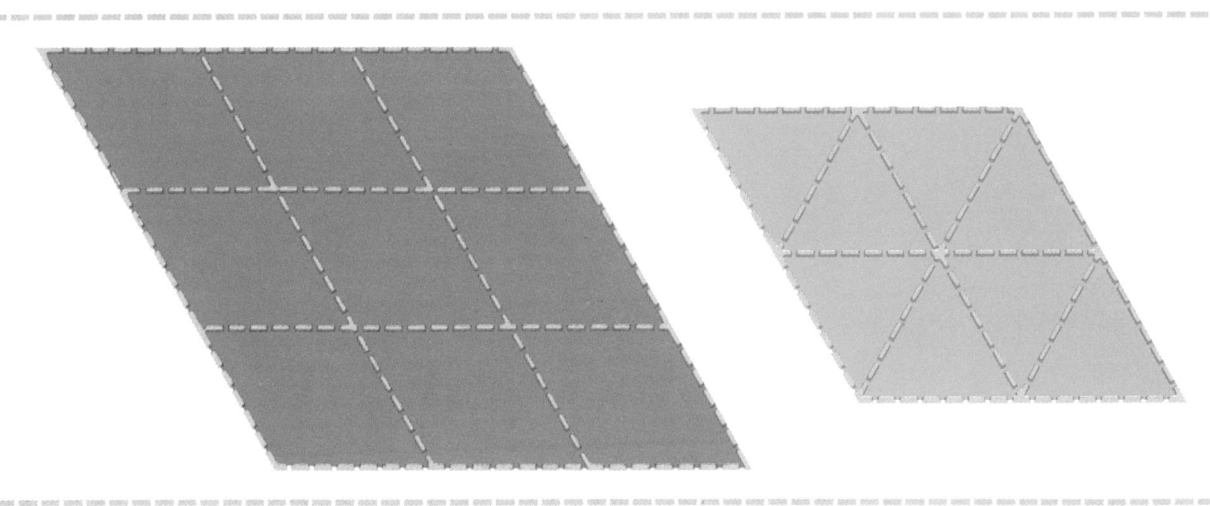

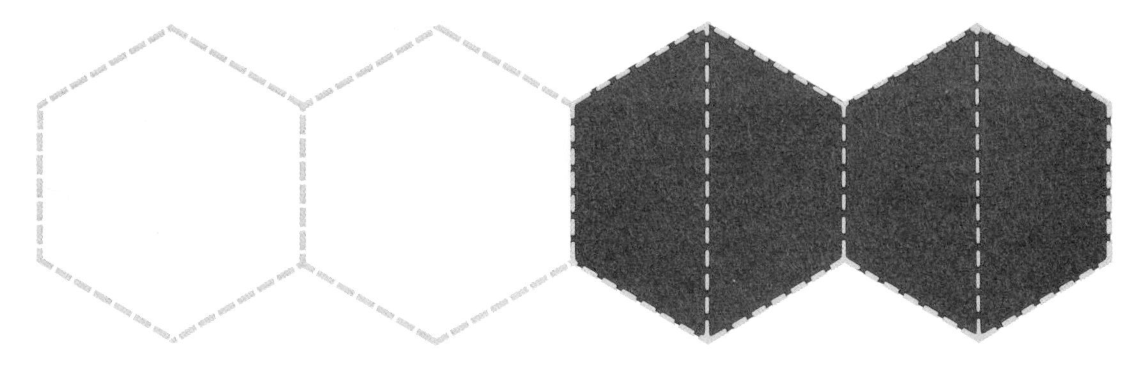

Use as described on page 76.

A Sunny Day

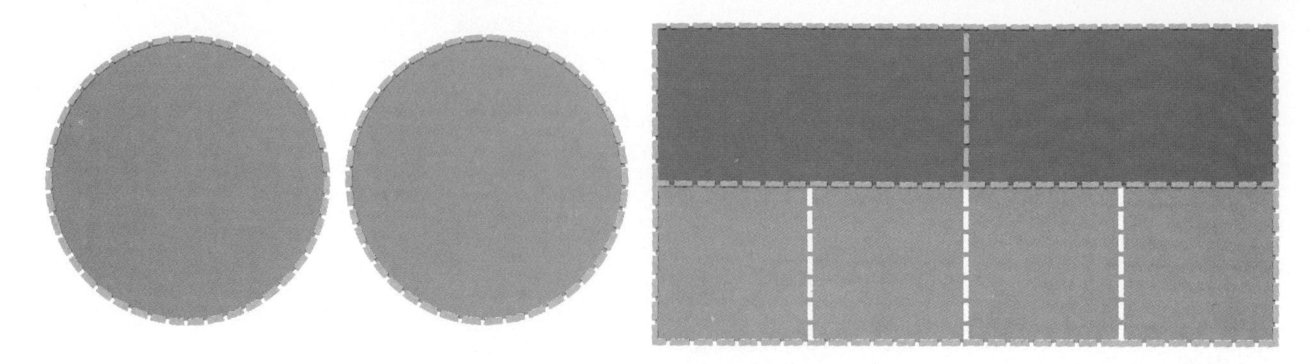

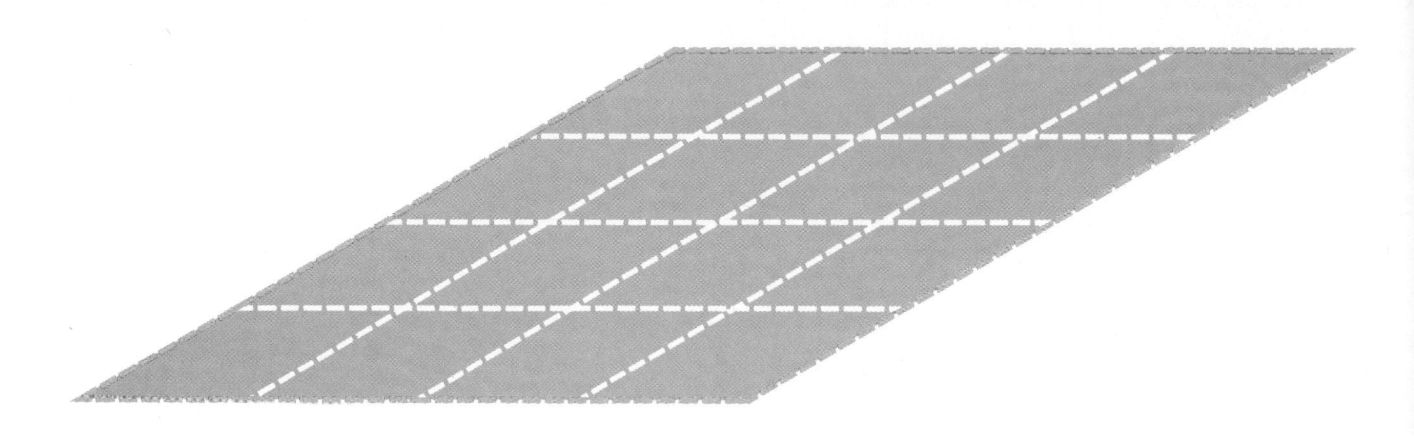

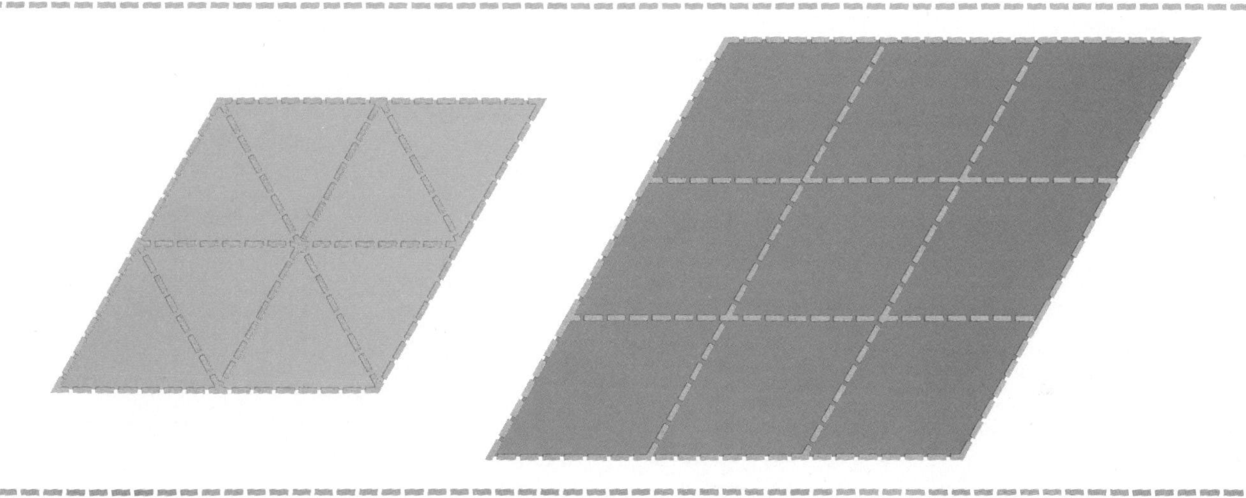

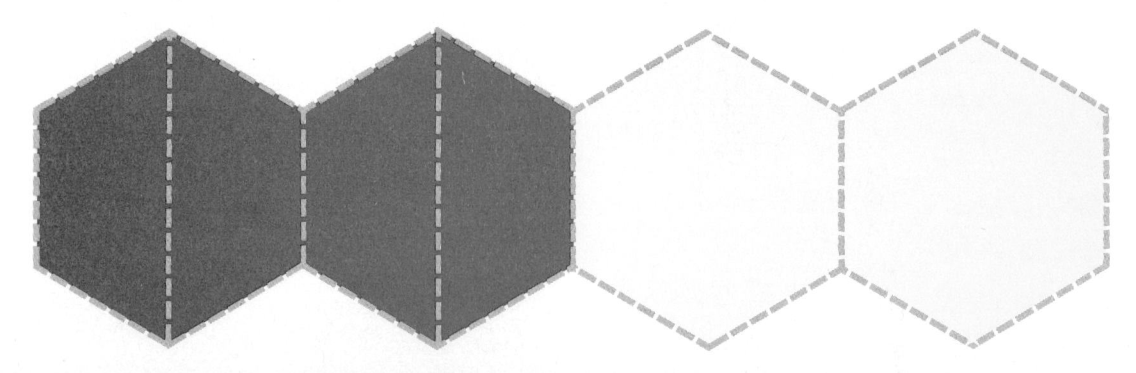

Use as described on page 76.

My first time.	My second time.

Use as described on page 76.

Ted's Wagon

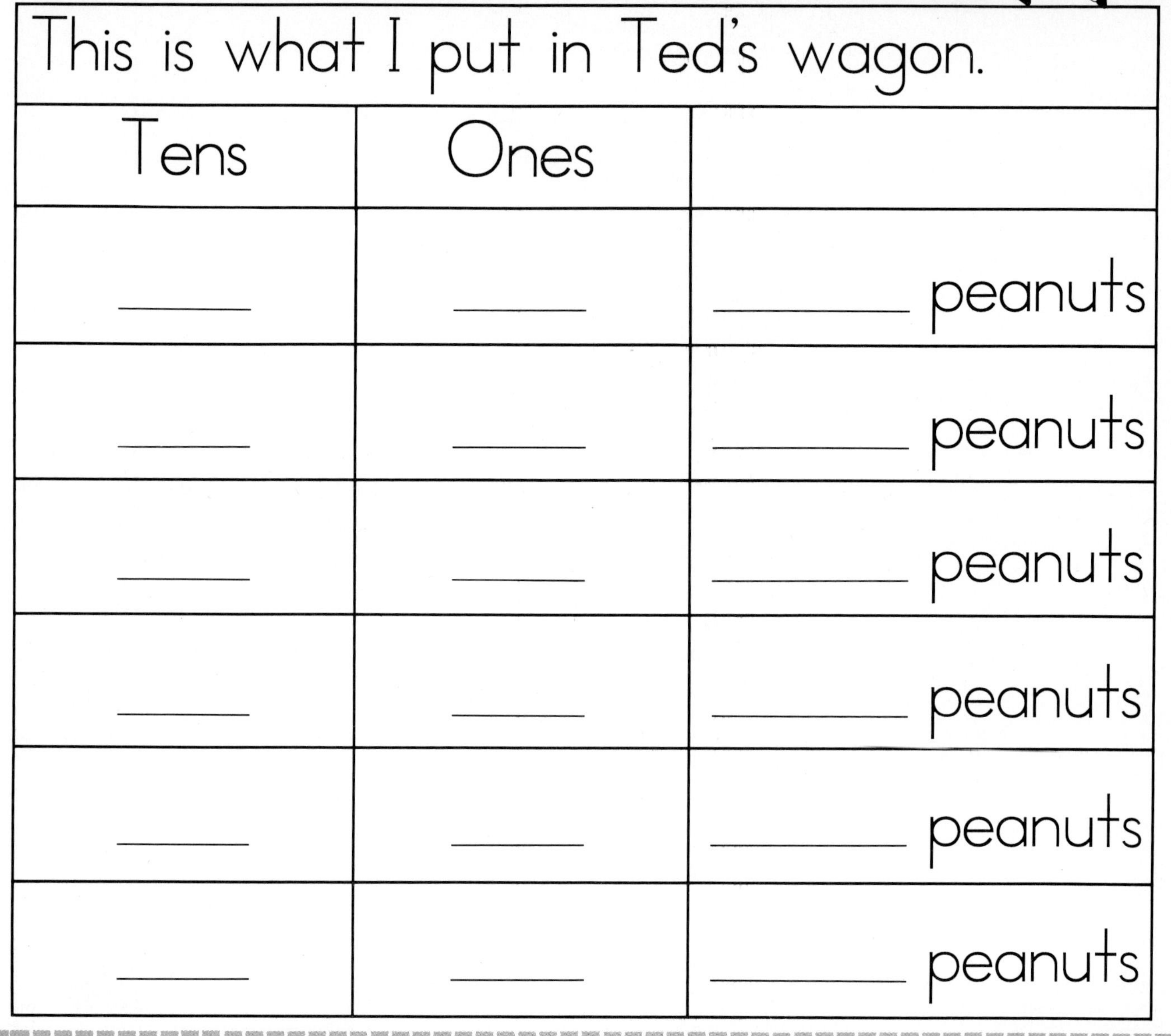

This is what I put in Ted's wagon.

Tens	Ones	
______	______	________ peanuts
______	______	________ peanuts
______	______	________ peanuts
______	______	________ peanuts
______	______	________ peanuts
______	______	________ peanuts

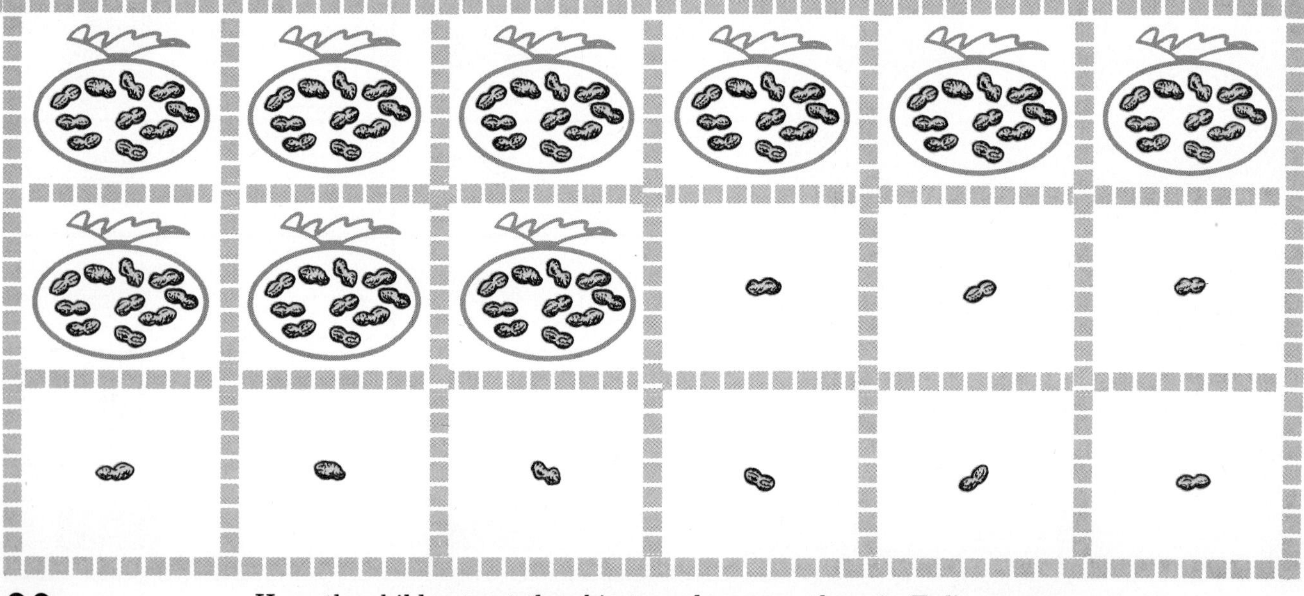

Have the child cut out the objects and arrange them in Ted's wagon on page 81 to form sets to 99. The sets can be recorded on this page. The child can glue down one load and print the numeral on the handle of the wagon.

Ted's Wagon

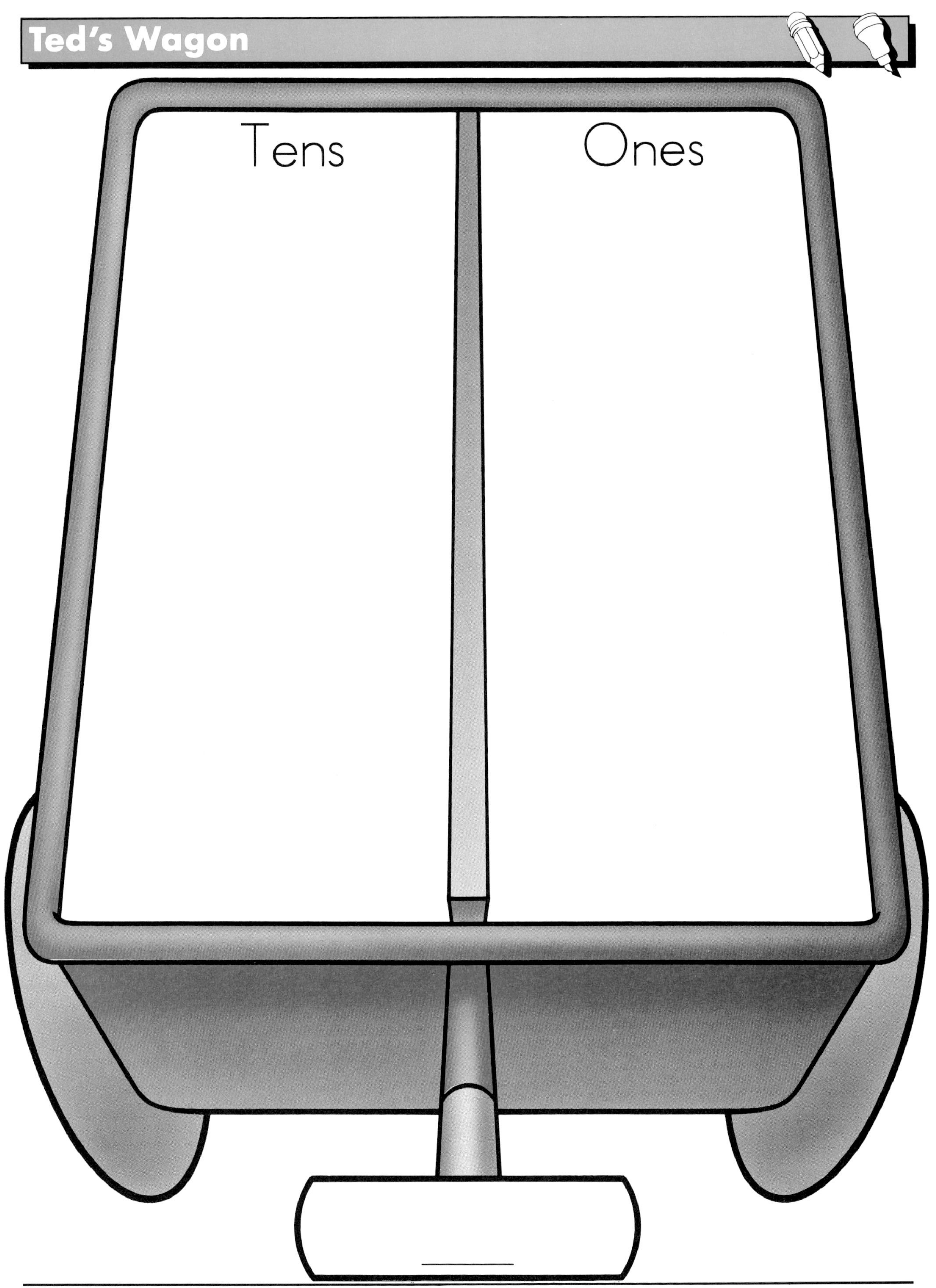

Use as described on the preceding page.

Ted's Shelves

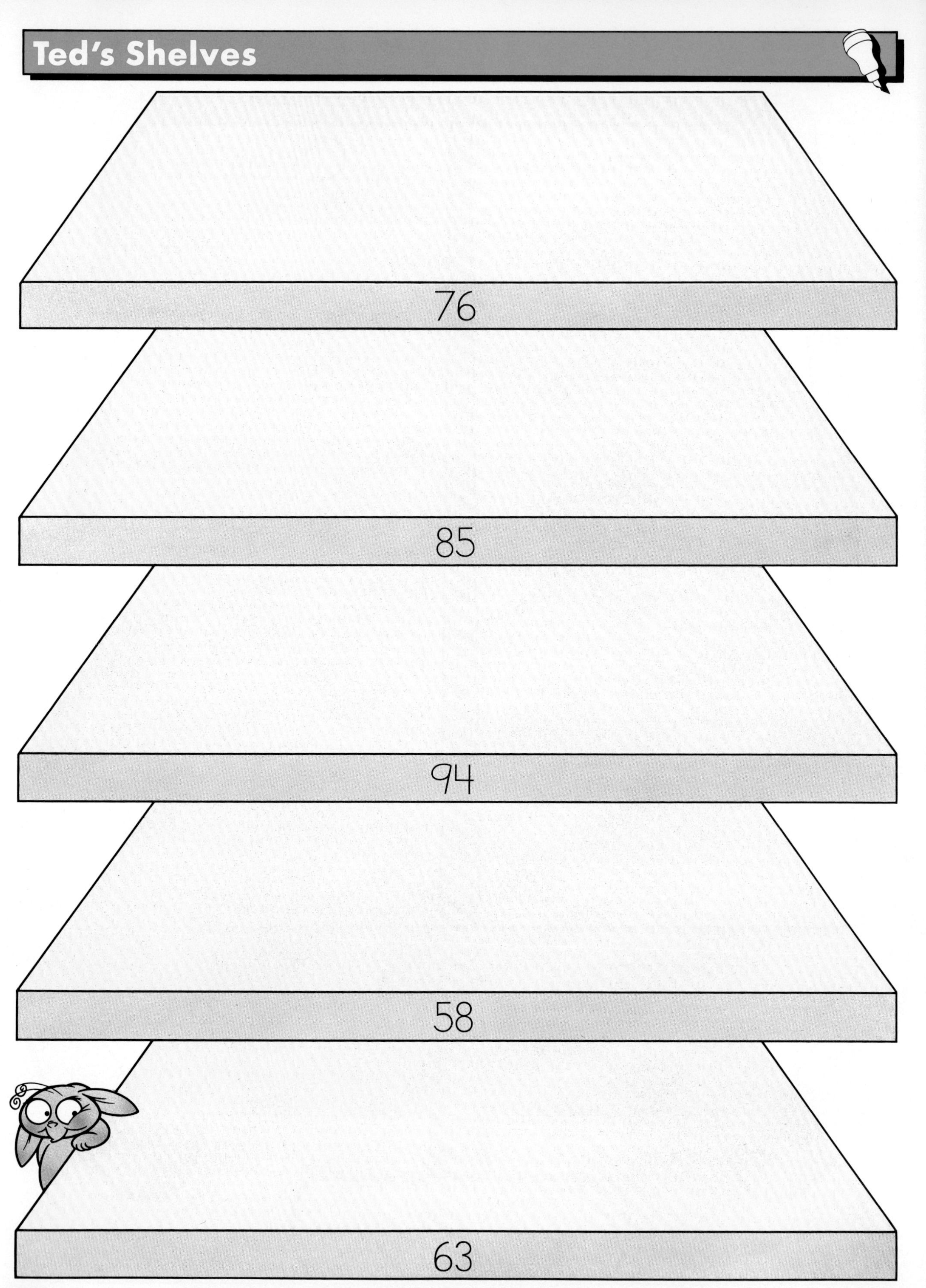

Have the child cut out and glue strips of items from page 83 or 84 so they match the numbers printed on these shelves.

Ted's Shelves

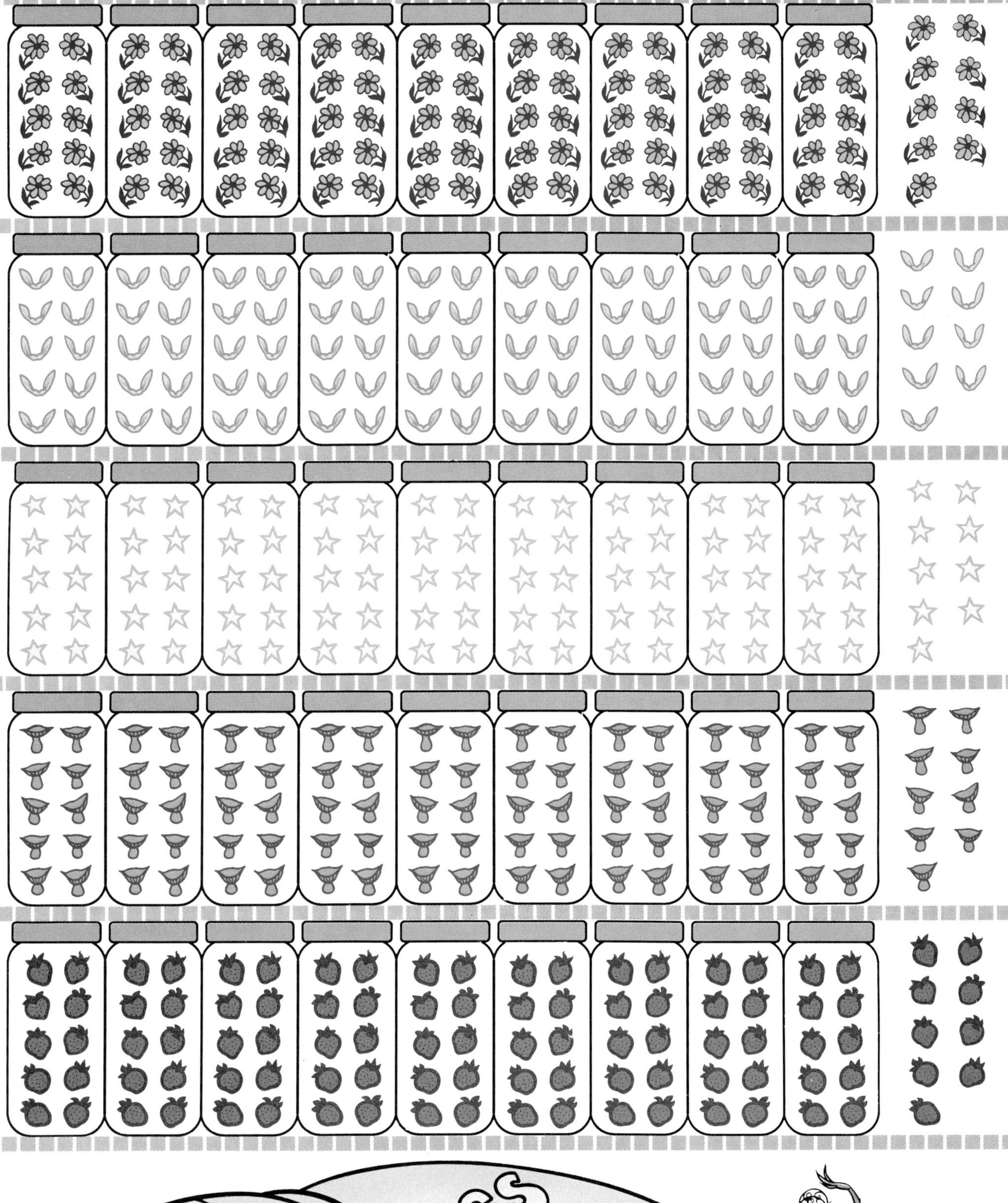

Have the child cut the strips so they represent the number identified on Ted's shelves on page 82. The extra pieces should be saved for use on page 85.

Ted's Shelves

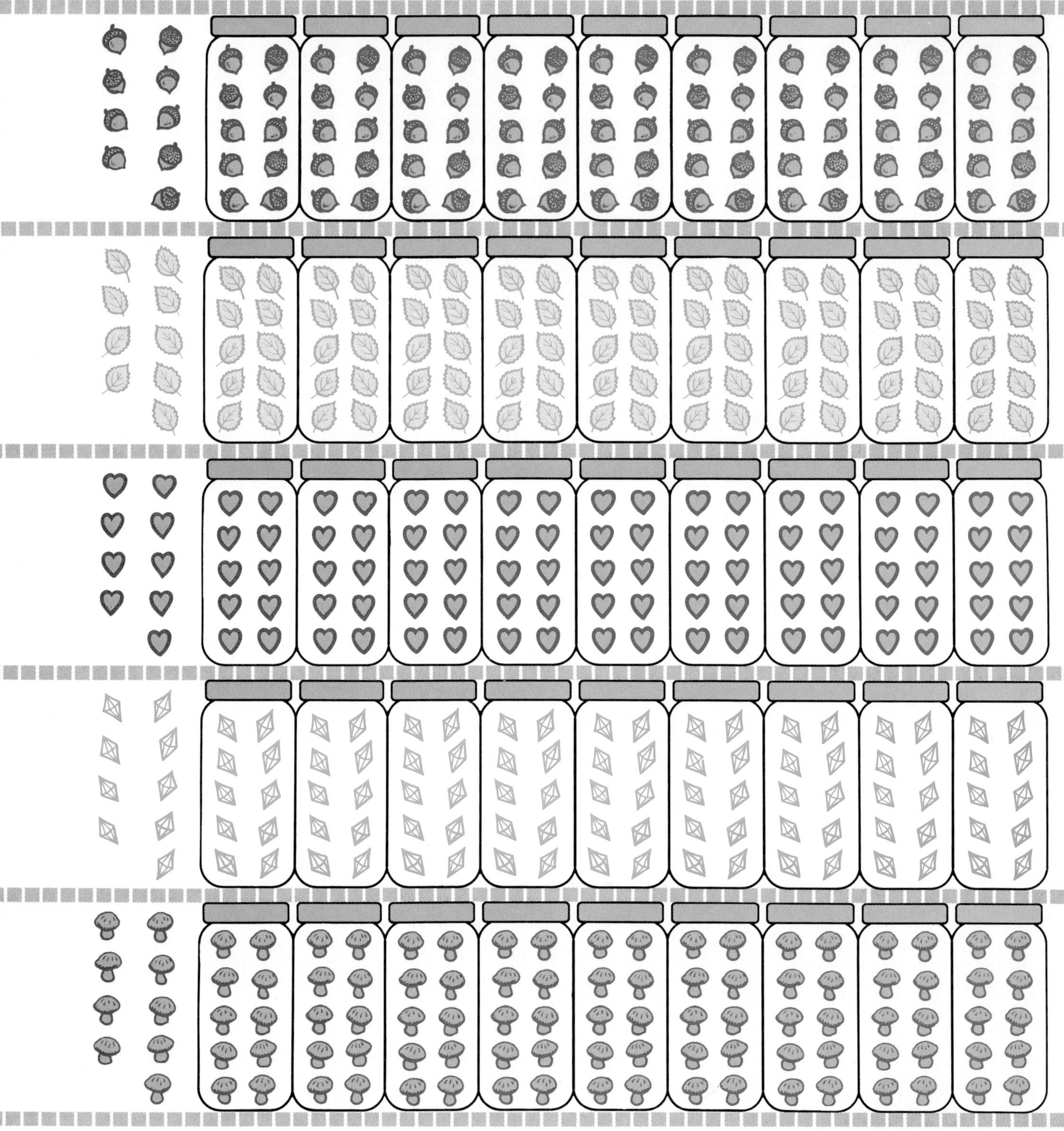

Use as described on the preceding page.

My Shelves

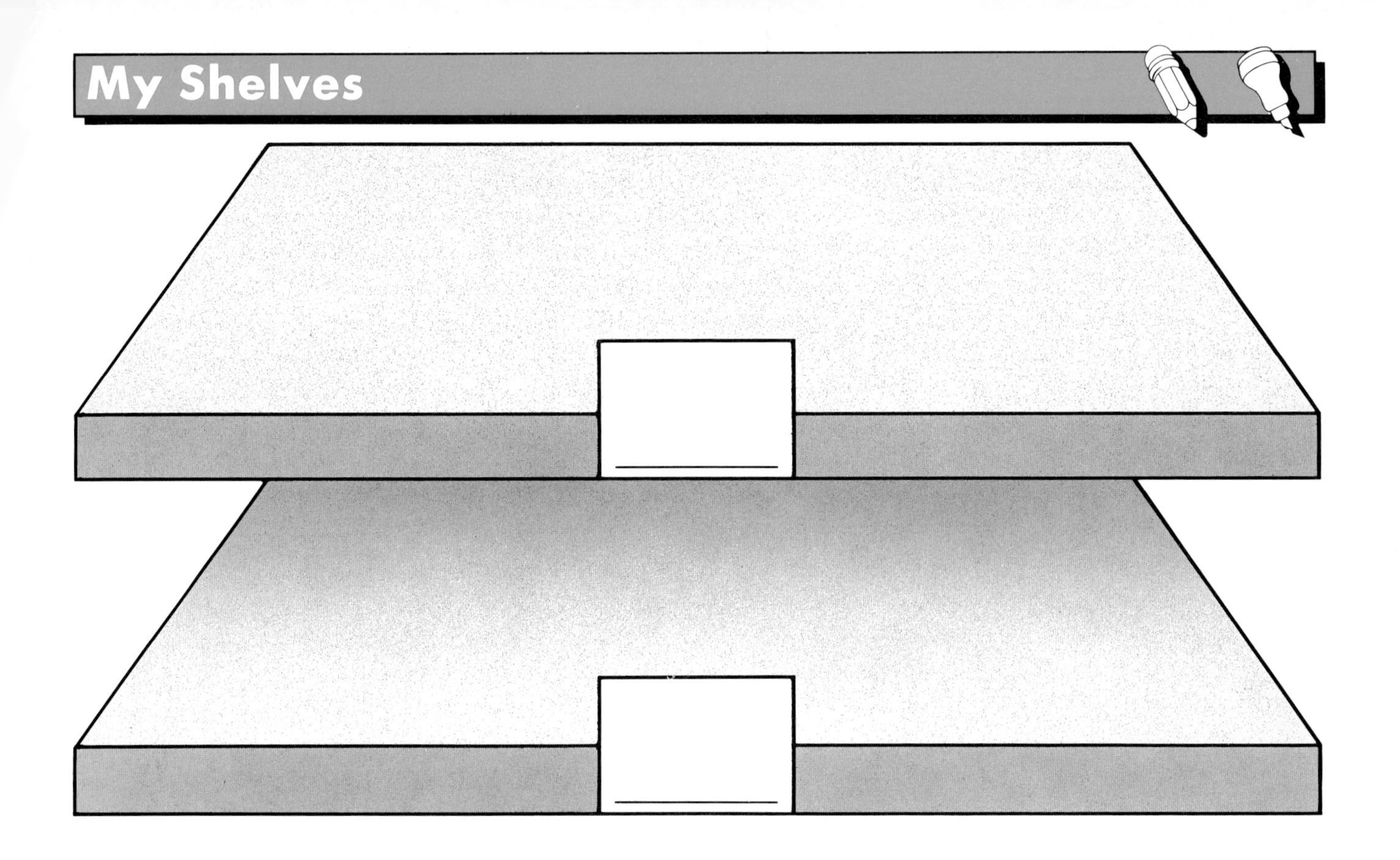

Have the child use the items remaining from pages 83 and 84 to create and label his or her own sets on the shelves provided.

Outdoor Fun!

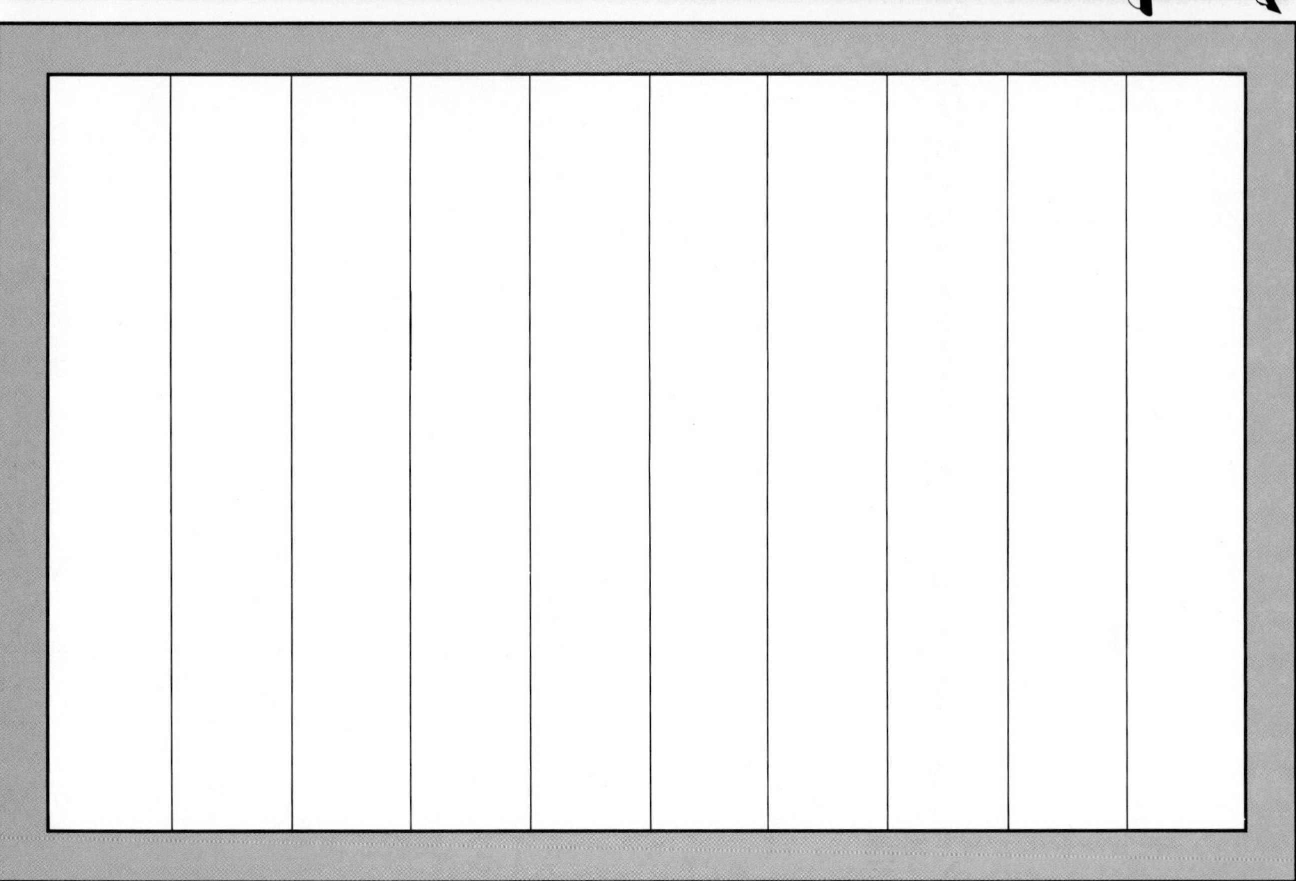

Have the child cut out the puzzle pieces and arrange them to form a picture at the top of the page.

Indoor Fun!

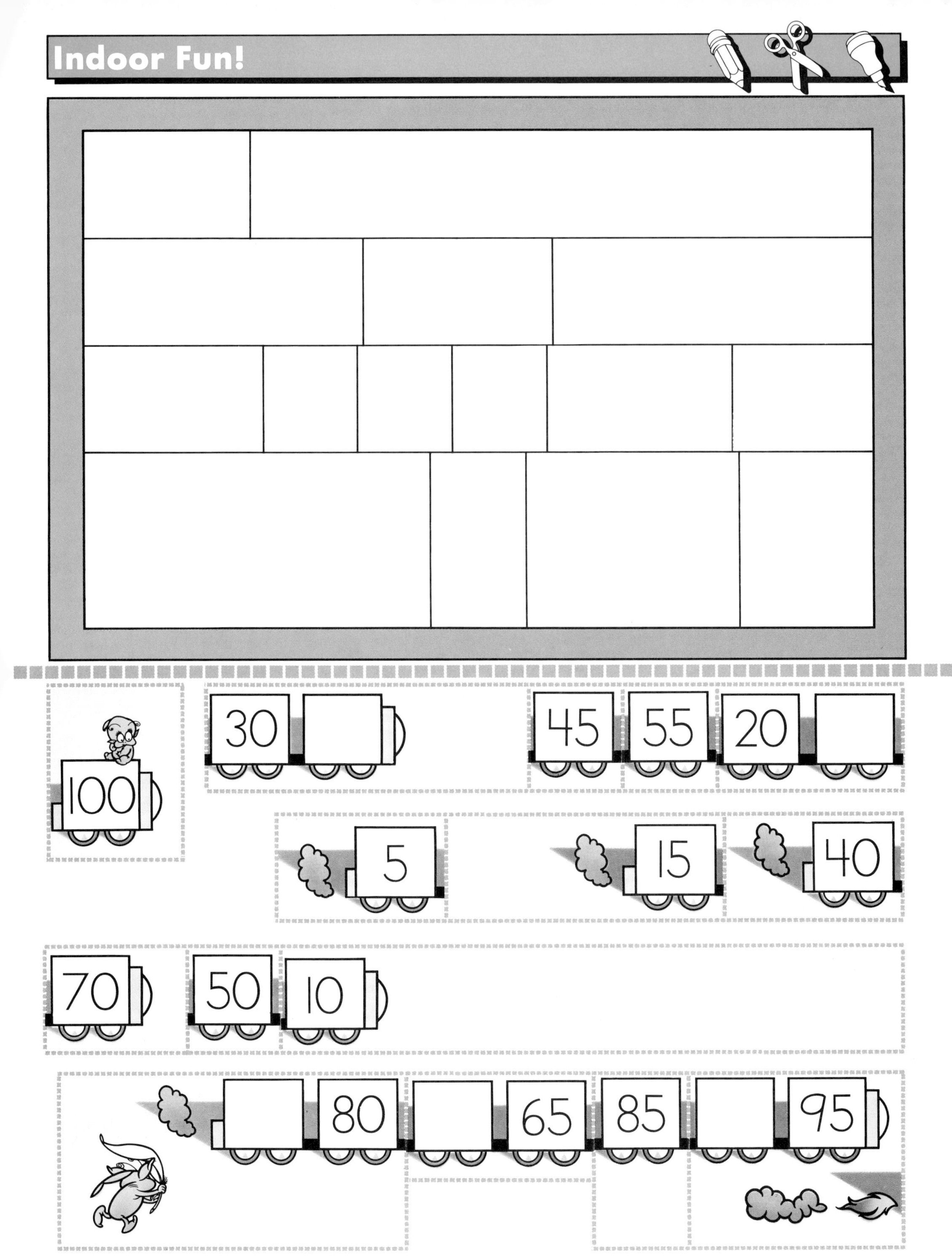

Have the child cut out the puzzle pieces and arrange them to form a picture at the top of the page. Encourage the child to count by fives to 100. Have the child record the missing numbers.

What's the Special Number?

1 2 3 4 5 6 7 8 9 10

11 12 13 14 15 16 17 18 19 20

The special number is:

- less than 8
- greater than 4
- say it counting by 2's

The number is: ______

The special number is:

- between 11 and 19
- less than 16
- greater than 12
- say it counting by 5's

The number is: ______

Have the child use the clues to find each special number.

Watch the Clock

Have the child cut out the digital clocks and glue each beside the matching alarm clock. Encourage the child to read the time aloud.

Ted's Time

Have the child print the time for each picture and describe an accompanying story.

Have the child print the time for each picture and describe an accompanying story.

Guess and Measure

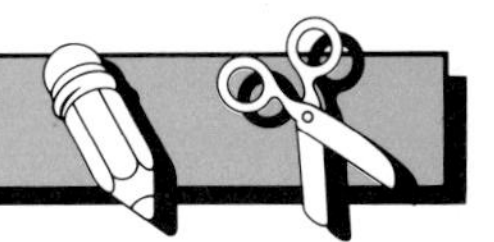

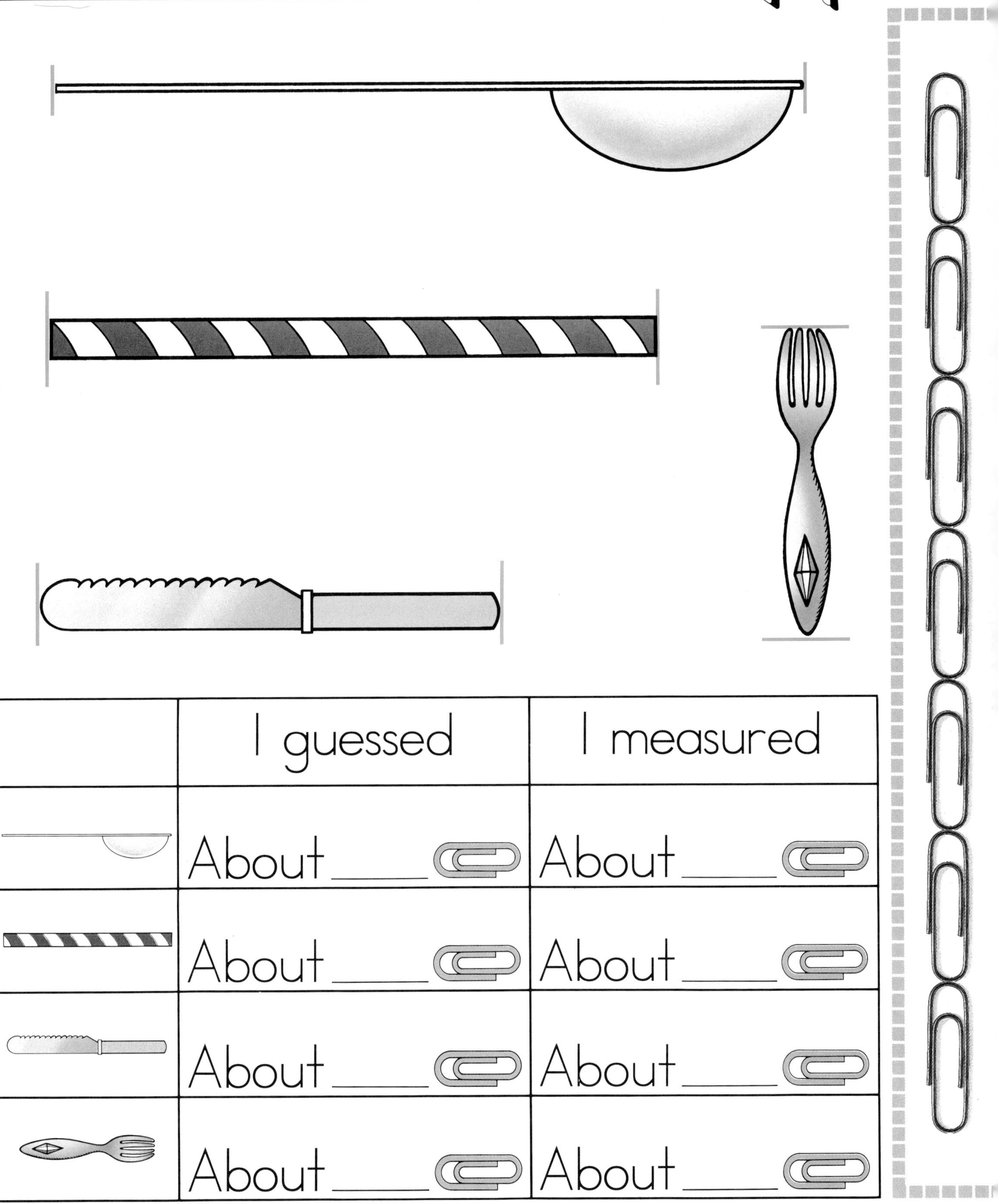

	I guessed	I measured
(spoon)	About ____	About ____
(straw)	About ____	About ____
(knife)	About ____	About ____
(fork)	About ____	About ____

Have the child estimate how long each item is before measuring with the cut-out units. Save the units for use on page 92.

Guess and Measure

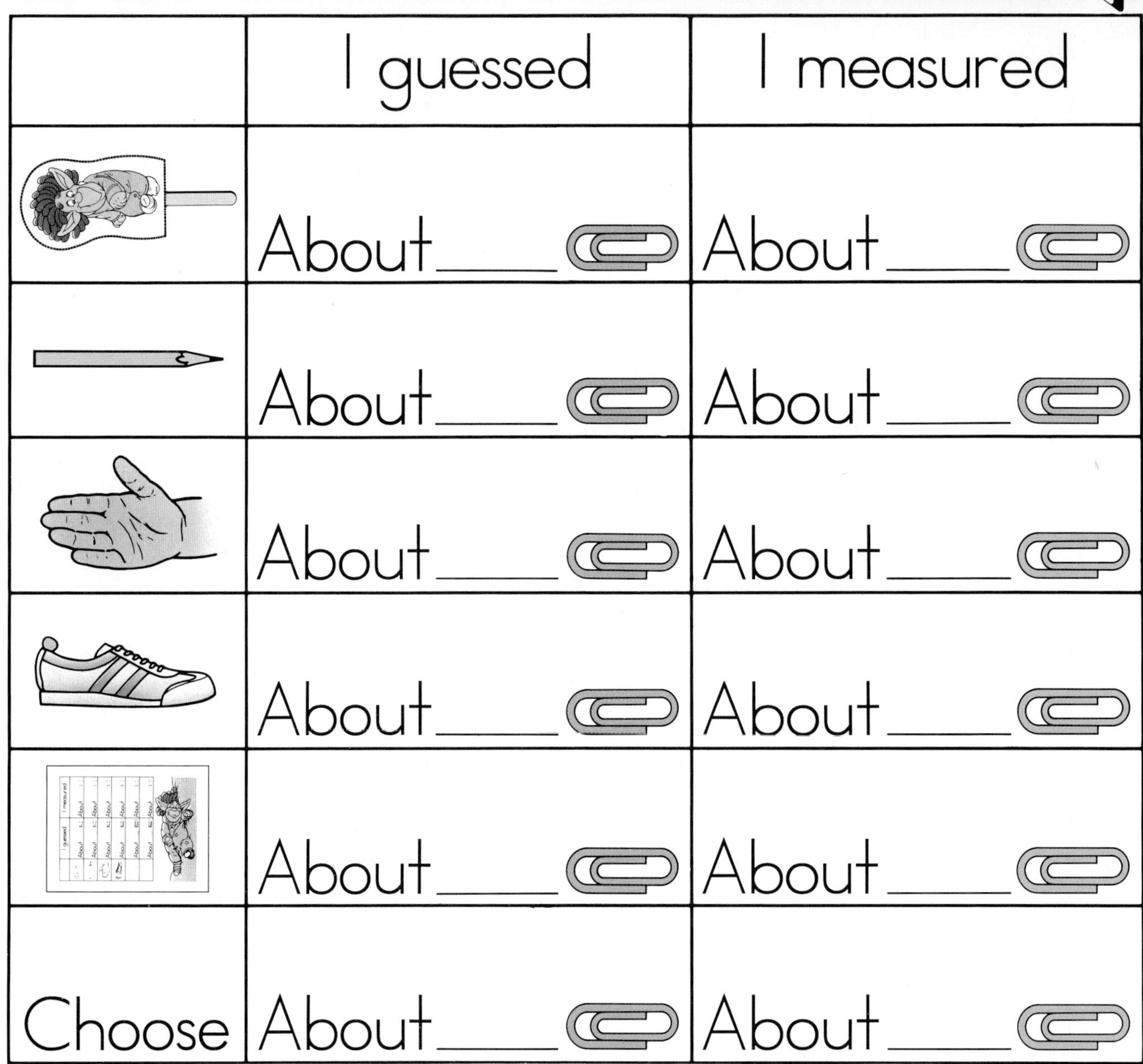

	I guessed	I measured
	About ____	About ____
	About ____	About ____
	About ____	About ____
	About ____	About ____
	About ____	About ____
Choose	About ____	About ____

Have the child estimate how long each item is before measuring with the cut-out units.

Puppets (pages 3-8) and Ted's Friends (pages 9 and 10)

Puppets can be used to:

- act out stories, songs, poems from *Explorations 1* and those known or created by the children.
- act out and solve story problems related to a variety of number concepts. For example, create sets of 0-10; create a set that is more, less or equal, or 1 more or 1 less than a given set; create combinations of a number, e.g., 4 and 0, 3 and 1, 2 and 2, 1 and 3, 0 and 4; choose the correct operation, e.g., *3 rabbits came to the meadow. 3 more animals came too. How many animals in the meadow? What is the number sentence?* $(3+3=6)$

All puppets can be used in plays while the finger puppets (pages 7 and 8) and Ted's friends (pages 9 and 10) can also be used on the story boards (pages 39-42 and 51-54). You may wish to store smaller puppets in an envelope.

Hundreds Charts (pages 37 and 56)

The hundreds chart can be used to:

- count by ones, twos, fives, and tens. Encourage the child to talk about the pattern.

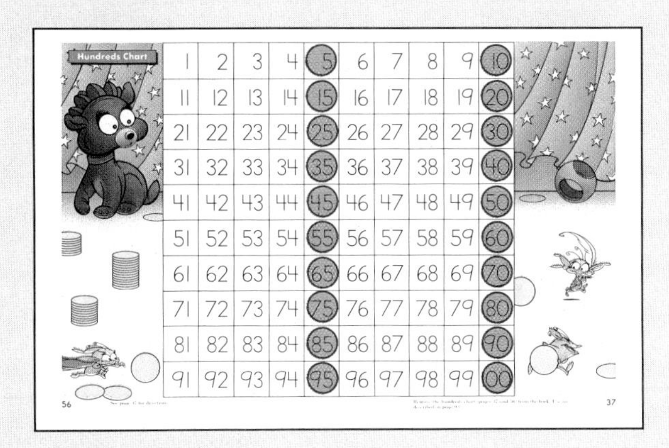

1	2	3	4	5	6	7	8	9	10
11	12	13	14	15	16	17	18	19	20
21	22	23	24	25	26	27	28	29	30
31	32	33	34	35	36	37	38	39	40
41	42	43	44	45	46	47	48	49	50
51	52	53	54	55	56	57	58	59	60
61	62	63	64	65	66	67	68	69	70
71	72	73	74	75	76	77	78	79	80
81	82	83	84	85	86	87	88	89	90
91	92	93	94	95	96	97	98	99	100

- count by tens beginning at any numeral on the top row, e.g., 7, 17, 27.... Talk about the pattern.
- cover even numbers to identify odd numbers and vice versa. Talk about the pattern.
- find and describe other counting patterns.
- count on or back 10 from a given number.
- identify the missing numerals when a row or a random selection have been covered.
- find mystery numbers, e.g., *the number is greater than 80, less than 95 and you say it when counting by tens. What is it?*

Graphing Mat (pages 38 and 55)

- Pose a question or have the child create a question related to a topic of interest. Begin by creating and talking about concrete graphs. Later have the child create pictographs.

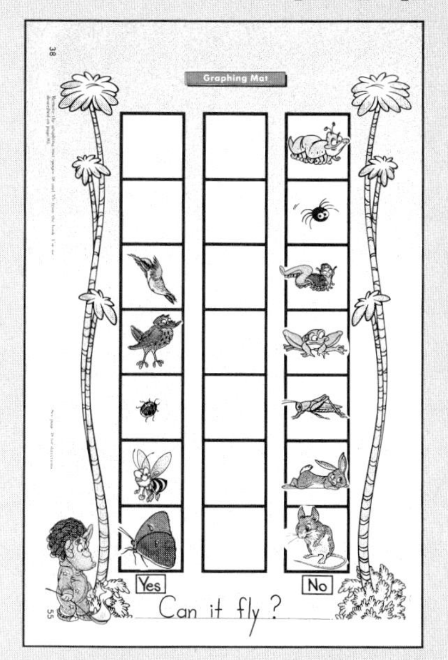

- Asking questions similar to these will assist the child in interpreting the graphs.
  - *Which column has more? less? or do they have the same number?*
  - *Are there more (less) __________ than __________?*
  - *What else can you tell me about your graph?* When interpreting a graph comparing 3 or more groups use the terms *most* and *least*.

Story Boards (pages 39-42 and 51-54)

Story boards may be used to reinforce a wide range of number skills. Have children use Ted's friends (pages 9 and 10) or finger puppets (pages 7 and 8) to act out story situations which involve any of the number concepts described.

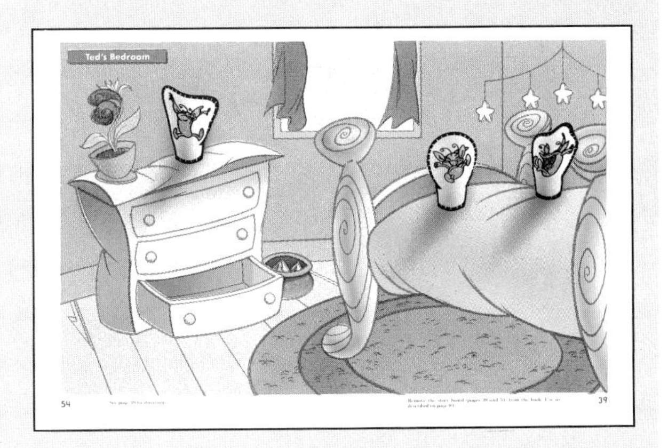

Ordering Mat (pages 43 and 50)

- Select an object. Place it on the mat with one end touching the base line. Ask the child to find objects that are longer and/or shorter, and arrange them on the mat according to length. Encourage the child to describe the order.
- Have the child select objects and arrange and describe them by length (or some other criterion).

Sorting Mat (pages 44 and 49)

Provide the child with a variety of materials to sort. Have the child:
- sort out a set of objects, e.g., all the paper things.
- sort everything in the group by one attribute, e.g., color, shape, size, texture.
- sort and resort the objects.

These questions may be asked:
- *How did you sort these?*
- *Why does (doesn't) this belong here?*
- *What other way could you sort these? Show me.*

Gameboards (pages 45 to 48) and Game Pieces (pages 69 and 70)

To form a large gameboard, have two children join their boards together, one using pages 45 and 48 and the other pages 46 and 47.

Many games can be created to reinforce number skills and concepts.

- Each child has a collection of 10 objects or imps (page 69). Numeral cards (page 69) are shuffled and placed face down. The player spins the color spinner. Before the marker is moved ahead, a numeral card is turned over. The task completed in response to the numeral can vary depending upon the skill you wish to reinforce. For example, the child could: create the appropriate set; a greater or lesser set; the set that comes before or after; create that set and add on to it until 10 has been formed; create a combination for that number, etc. If the other players agree that the task has been completed, the child moves the marker to the next colored space that corresponds to the spinner.
- A set of numeral cards (page 69) are placed down. The top card is turned over and will be the 'game card'. A child spins the color spinner. Before the marker is moved ahead, the child turns over a numeral card and completes a task, for example, identifies if his or her number is greater or less than the game card; creates and solves an addition or subtraction sentence using the two numbers; counts on or back from the game card to his or her own number, etc.
- The procedure above could be adapted for work with place value. For example, establish that the game card is a tens card and the numeral card turned over for each turn is a ones card. The child could form a 2-digit number with the 2 cards and complete a task such as the following before moving a marker forward:
  —identify the number and create the set;
  —identify the number before and after;
  —count on or back 10 numbers; etc.
- Children can use the numeral and +, − and = and a set of the 2-sided imp cards (page 69) to play Bouncing on the Bed (page 71) in a game situation.
- Have the children create their own games.

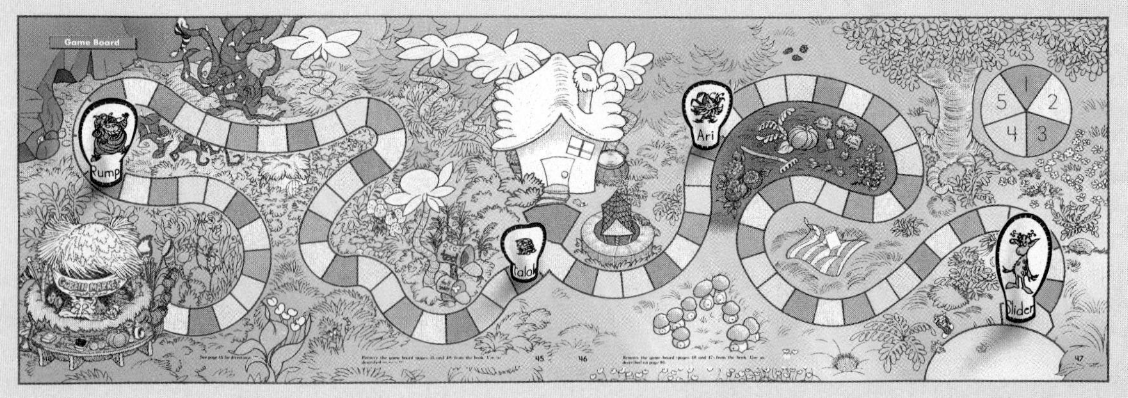